CANCER STEM CELLS

A remarkable paradigm shift has occurred in recent years regarding the biological origins of cancer. The cancer stem cell hypothesis has challenged the foundational notions of cancer, and the therapeutic implications have been profound. Compelling evidence indicates that errors in the development of a small subset of adult stem cells can lead to cancer. Only this small subpopulation of cells has the inherent ability to form tumors and metastasize. This book discusses the emerging field of cancer stem cell research, with contributions from leading experts on the basic biology, genetic pathways, and potentials for therapeutic targeting of cancer stem cells. It also covers clinical challenges for these new discoveries, namely, that cancer stem cells might be resistant to conventional chemotherapeutic and radiological treatments and may be at the biological core of relapse and therapeutic resistance. This book is an essential concise guide to the latest discoveries and therapies in cancer research.

William L. Farrar, PhD, is head of the Cancer Stem Cell Section at the Laboratory of Cancer Prevention, National Cancer Institute, Frederick, Maryland.

Cancer Stem Cells

Edited by

William L. Farrar

National Cancer Institute, Frederick, Maryland

CAMBRIDGE
UNIVERSITY PRESS

CAMBRIDGE
UNIVERSITY PRESS

Shaftesbury Road, Cambridge CB2 8EA, United Kingdom

One Liberty Plaza, 20th Floor, New York, NY 10006, USA

477 Williamstown Road, Port Melbourne, VIC 3207, Australia

314–321, 3rd Floor, Plot 3, Splendor Forum, Jasola District Centre, New Delhi – 110025, India

103 Penang Road, #05-06/07, Visioncrest Commercial, Singapore 238467

Cambridge University Press is part of Cambridge University Press & Assessment, a department of the University of Cambridge.

We share the University's mission to contribute to society through the pursuit of education, learning and research at the highest international levels of excellence.

www.cambridge.org
Information on this title: www.cambridge.org/9780521896283

© Cambridge University Press & Assessment 2010

First published 2010

A catalogue record for this publication is available from the British Library

Library of Congress Cataloging-in-Publication data
Cancer stem cells / edited by William L. Farrar.
    p. ; cm.
Includes bibliographical references and index.
ISBN 978-0-521-89628-3 (hardback)
1. Cancer cells.   2. Stem cells.   I. Farrar, William L.   II. Title.
[DNLM: 1. Neoplastic Stem Cells – physiology.   2. Neoplasms – etiology.
3. Signal Transduction – physiology. QZ 202 C215657 2009]
RC269.7.C375   2009
616.99´4027 – dc22   2009010915

ISBN   978-0-521-89628-3   Hardback

Contents

Color plates follow page 78.

Contributors

Fritz Aberger, PhD
Department of Molecular Biology
University of Salzburg
Salzburg, Austria

Massimiliano Bonafe, MD
Center for Applied Biomedical
  Research
St. Orsola-Malpighi University
  Hospital
Bologna, Italy
Department of Experimental
  Pathology
University of Bologna
Bologna, Italy

Stephen Byers, PhD
Departments of Oncology,
  Biochemistry, Molecular and
  Cellular Biology, and Surgery
Lombardi Comprehensive Cancer
  Center
Georgetown University
Washington, DC

William L. Farrar, PhD (EDITOR)
Cancer Stem Cell Section
Laboratory of Cancer Prevention
Center for Cancer Research
National Cancer Institute
National Institutes of Health
Frederick, Maryland

Robert Glazer, PhD
Departments of Oncology,
  Biochemistry, Molecular and
  Cellular Biology, and Surgery
Lombardi Comprehensive Cancer
  Center
Georgetown University
Washington, DC

Gennadi Glinsky, MD, PhD
Ordway Research Institute
Albany, New York

Monica L. Guzman, PhD
University of Rochester School of
  Medicine and Dentistry
Rochester, New York

Meenhard Herlyn, DVM, DSc
Program of Molecular and Cellular
  Oncogenesis
The Wistar Institute
Philadelphia, Pennsylvania

Elaine M. Hurt, PhD
Cancer Stem Cell Section
Laboratory of Cancer Prevention
Center for Cancer Research
National Cancer Institute
National Institutes of Health
Frederick, Maryland

Collene R. Jeter, PhD
Department of Carcinogenesis
The University of Texas M. D.
   Anderson Cancer Center
Smithville, Texas

Craig T. Jordan, PhD
Departments of Medicine and
   Biomedical Genetics
University of Rochester Medical
   Center
Rochester, New York

Lopa Mishra, PhD
Departments of Oncology,
   Biochemistry, Molecular and
   Cellular Biology, and Surgery
Lombardi Comprehensive Cancer
   Center
Georgetown University
Washington, DC

Joon T. Park, PhD
Departments of Pathology and
   Obstetrics/Gynecology
Johns Hopkins Medical Institutions
Baltimore, Maryland

Michael Pishvaian, PhD
Departments of Oncology,
   Biochemistry, Molecular and
   Cellular Biology, and Surgery
Lombardi Comprehensive Cancer
   Center
Georgetown University
Washington, DC

Alexander Roesch, PhD
Program of Molecular and Cellular
   Oncogenesis
The Wistar Institute
Philadelphia, Pennsylvania

Gerrit J. Schuurhuis, PhD
Department of Hematology
VU University Medical Center
Amsterdam, The Netherlands

Stewart Sell, MD
Wadsworth Center
Ordway Research Institute
University at Albany
Albany, New York

Ie-Ming Shih, PhD
Department of Pathology
Johns Hopkins Medical Institutions
Baltimore, Maryland

Dean G. Tang, MD, PhD
Department of Carcinogenesis
University of Texas M. D. Anderson
   Cancer Center
Smithville, Texas

Tian-Li Wang, PhD
Department of Obstetrics/Gynecology
Johns Hopkins Medical Institutions
Baltimore, Maryland

Preface

Cancer results from the accumulated effects of somatic or inherited gene alterations that result in the improper function of proteins. An increased understanding of the underlying genetics has shaped the modern hypotheses for the basis of cancer. First was the concept of *oncogenes*, defined as genes that promote a transformed cellular phenotype. The altered activities of this class of proteins are usually due to mutations in the genes themselves, polymorphisms in promoter elements, or aberrant activation of upstream signaling pathways. The next concept with profound implications for the genetic basis of cancer was the discovery of tumor suppressor genes. This class of genes, when genetically silent, essentially takes the brakes off the normal controls of cell cycle, senescence, and apoptosis. From the silencing of genes in cancer emerged the rapidly growing field of epigenetics and how gene silencing leads to the development of cancer. Therefore, for the past several decades of molecular biology, the focus has been on the ON and OFF switching of genes. Engineering of recombinant DNA in model cell systems produced a greater understanding of the underlying biochemistry and molecular biology of cancer. This led to the belief that similar alterations could occur naturally in nearly any somatic cell type, and therefore cancer was believed to be of a stochastic nature.

The *stochastic hypothesis* suggests the clonal evolution model, in which any cell with overexpressed oncogenes and/or downregulated tumor suppressors will eventually form a tumor. This model could explain the multiple aspects of human disease and clinical observations. However, recently, a hypothesis has reemerged to challenge this notion, causing a shift away from the stochastic model. Increasing evidence, initially discovered in hematological malignancies and, later, in solid tumors, suggests that tumors are formed from a subset of cells with unique characteristics that reside within the volume of the tumor. The unique subset of tumor-initiating cells is defined as *cancer stem cells*, a term initially coined by researchers in hematological malignancies and adopted by solid tumor researchers. What is shared in common with diverse cancers is that the unique subsets of tumor-initiating cells have stem cell–like biological and genetic similarities. Most pronounced are unique sets of surface markers, the ability of self-renewal, expression of developmental stem cell–like genes, and biological properties that facilitate tumor development.

The birth of the cancer stem cell hypothesis has generated a large degree of enthusiasm not without profound therapeutic considerations. For the most part, few of the current chemotherapeutic and irradiation strategies have considered the cancer stem cell component of the tumor burden. In fact, there are significant indications that the tumor-initiating cells are resistant to the conventional tools of cancer therapeutics.

This book focuses on the clinical and therapeutic implications of cancer stem cells. We have included chapters concerning the basic science of both leukemic and solid tumor stem cell biology and a practical chapter on the isolation and characterization of cancer stem cells. Because of the initial recognition of cancer stem cells in leukemia, therapeutic strategies may first be employed in this cancer, as discussed by researchers active in the field. Finally, we have included chapters describing stem cell signaling pathways that direct self-renewal and other vital cancer stem cell characteristics. These pathways offer the fodder for molecularly targeted therapeutics and rational drug design.

While this is a rapidly emerging field, the discovery of the cancer stem cell as a subset of cells with unique biological and genetic properties will likely have a substantial impact on cancer therapeutics and prevention as well as on the understanding of the biological origins of cancer.

William L. Farrar, PhD

SECTION I: CHARACTERIZATION OF CANCER STEM CELLS

1 Purification and characterization of cancer stem cells

Elaine M. Hurt and William L. Farrar

National Cancer Institute at Frederick, National Institutes of Health

The processes underlying the etiology of cancer have been the fodder for several theories for a century (for a discussion of the earliest theories, see the subsequent discussion and previous studies).[1,2] Central to all these theories is the cell of origin for the transformation from a normal to a cancerous cell. The prevailing hypothesis, until recent years, was that any cell that had acquired multiple genetic hits could give rise to a tumor.[3] The cancer stem cell hypothesis posits that only a small subset of cells, termed *tumor-initiating cells* or *cancer stem cells* (CSCs), is capable of giving rise to and maintaining tumors. Therefore all CSCs must display several characteristics: they must be the only cells that are capable of giving rise to a tumor (*tumorigenic*), they must be able to maintain the population of tumorigenic cells (*self-renewal*), and they must be able to give rise to the heterogeneous cells composing the entire tumor (*pluripotency*). When a CSC is transplanted into an immunocompromised mouse, self-renewal and pluripotency are vital for

The content of this publication does not necessarily reflect the views or policies of the Department of Health and Human Services, nor does mention of trade names, commercial products, or organizations imply endorsement by the US Government.

the formation of a tumor that recapitulates the original (reviewed by Wang and Dick[4]).

HISTORY OF CANCER STEM CELLS (CSCs)

Tumors are masses containing heterogeneous populations of cells with different biological characteristics.[5] Although there has been a marked increase in the number of publications regarding CSCs in the past 14 years, the notion that cancer cells have properties reminiscent of stem cells is not a new theory. This idea was first postulated by Rudolph Virchow and Julius Cohnheim in the nineteenth century.[1,6] Virchow's *embryonal rest hypothesis* noted the similarities between fetal tissue and cancer cells with respect to their ability to proliferate and differentiate.[2] Later, Cohnheim and Durante extended this by hypothesizing that there exist embryonal remnants in mature organs, and Beard hypothesized that cancer arises either from activated germ cells or from dislodged placental tissue. Within all these hypotheses, the basis for cancer was a cell that maintained the ability to differentiate, renew, and proliferate in a manner similar to cells of the developing embryo.

The first demonstration that tumors comprise cells with differential tumor-forming ability was in 1961. Southam and Brunschwig harvested recurrent cancer cells from patients and then autotransplanted the cells into different sites. To establish a new tumor, at least one million cells needed to be injected, and this only worked approximately 50% of the time.[7] Later studies showed similar results for colony initiation in vitro.[8,9] This suggested that not all the cells could initiate a tumor and that there existed a hierarchy of tumor-initiating cells.

The first demonstration of the hierarchy of cancer cells was done in leukemia by Lapidot and colleagues.[10] They demonstrated that CD34+CD38− cells isolated from acute myeloid leukemia patients developed a tumor when injected into nonobese diabetic/severe combined immunodeficiency (NOD/SCID) mice, but injection of even larger numbers of the more differentiated cells, CD34+CD38+, did not initiate tumor formation. Moreover, the tumors formed by injection of the CD34+CD38− cells were similar in morphology to the original disease present in the patient. Following leukemia, the first identification of CSCs in solid tumors was demonstrated in breast cancer by al-Hajj and colleagues in 2003.[11] Since then, CSCs have been identified in many solid tumors, including brain, prostate, pancreatic, liver, colon, head and neck, lung, and skin tumors.

IDENTIFICATION OF CSCs

Three methods are commonly employed for the isolation of CSCs. These methods include (1) the isolation by flow cytometric sorting of a side population (SP) based on Hoechst dye efflux, (2) sorting on the basis of cell surface marker expression, and (3) sphere culture. These methods lead to varying degrees of enrichment of CSCs, and each has its advantages and limitations.

Side populations

It was the observation of Goodell and colleagues that there was a small population of cells in bone marrow aspirates that did not accumulate Hoechst 33342 dye.[13] They further showed that this SP contained cells capable of repopulating the bone marrow. Using flow cytometry, the SP has been isolated from a variety of tumors, including leukemia,[14,15] ovarian cancer,[16] hepatocellular carcinoma,[17] brain cancer,[18-20] lung cancer,[21] thyroid cancer,[22] nasopharyngeal carcinoma,[23] prostate cancer,[24] breast cancer,[24] and other cancers. The SPs of all these tumor types have been shown to enrich cells with stemlike characteristics such as increased tumorigenicity,[14,16,17,19,21,23,24] expression of stemlike genes,[17,21-24] and self-renewal.[17,19,21,22,24]

It is generally thought that the SP is the result of the dye being extruded out of the cell by an ATP-binding cassette (ABC) transporter.[25] Indeed, bone marrow cells isolated from $abcg2^{-/-}$ mice lack an SP,[26] strong evidence that the bone marrow SP is a result of the efflux of Hoechst dye mainly by ABCG2. Moreover, the SP of neuroblastomas had increased expression of ABCG2 and ABCA3,[18] and the SP isolated from the breast cancer cell line, MCF7, has been studied extensively and has increased expression of ABCG2.[24,27] However, expression of ABCG2 alone may not identify the CSCs in all tumor types. In prostate cancer, the SP enriched tumor-initiating cells and ABCG2 expression, but purified ABCG2$^+$ cells did not show increased tumorigenesis compared to ABCG2$^-$ cells.[24] Thus the authors concluded that the SP is enriched for CSCs and that the SP contains the ABCG2$^+$ cells, but it is still a heterogeneous population, and ABCG2$^+$ cells are not the tumor-initiating cells. The expression of multidrug resistance 1 (MDR1), another drug transporter, is also not correlated with the SP in acute myeloid leukemia.[15] Therefore the lack of Hoechst staining in the SP may not rely entirely on efflux by drug transporters.

There is some evidence that suggests that the presence of the SP may be a result of inefficient dye uptake as a reflection of the presence of largely quiescent cells, another characteristic of stem cells. In prostate cancer, Bhatt and colleagues demonstrated that the CSC population was composed of the $G_{(0)}$ cells contained within the SP, whereas the $G_{(1)}$ cycling cells in the SP were the more differentiated transit-amplifying cells.[28] Likewise, Ho and colleagues showed that lung cancer SP cells were also largely quiescent.[21] This might, in part, explain the results of Patrawala and colleagues, which showed that the SP displayed increased tumorigenicity, but the ABCG2$^+$ population did not.[24] It may be that the CSCs, which are generally quiescent, are unstained by the Hoechst dye rather than actively transporting the dye out via a transporter.

Although the SP has been shown in many tissue types to enrich for CSCs,[29] it is generally agreed that it does not represent a homogeneous population of CSCs. Furthermore, in some cases, such as in nontransformed renal cells[30] and skin cells,[31,32] the SP does not appear to enrich cells with stem cell characteristics. Further limitations of this method of isolation have to do with the procedure itself, in which the parameters of Hoechst 33342 concentration, staining time, and staining temperature are critical. An excellent protocol can be found online (http://www.bcm.edu/labs/goodell/protocols.html). However, dye concentrations and staining time can vary with different cell preparations, and Hoechst staining needs to be carefully controlled every time it is performed.[29] Moreover, there have been reports that the dye can have deleterious effects on cells. For example, in the rat C6 glioma cell line, Shen and colleagues demonstrated that incubation with Hoechst 33342 for prolonged periods of time leads to increased apoptosis.[20] This problem raises the possibility that differences seen in tumorigenicity between SP and non-SP cells may be due to a toxic effect of the Hoechst dye, specifically in the non-SP population.

Cell surface markers

Cell surface markers have been used as a means of identification and isolation. Most of the markers utilized to date are based on knowledge of tissue development or are derived from hematopoietic or embryonic stem cells. The two most commonly used surface markers used to identify CSCs are CD133 and CD44.

Prominin-1 (CD133) was originally identified on rat neuroepithelial stem cells[33] in 1997. Later that year, a monoclonal antibody (AC133) was made to CD34$^+$ stem cells isolated from fetal liver, bone marrow, and cord blood,[34] and subsequent cloning identified it as the human homolog of prominin-1. Prominin-1 is a transmembrane glycoprotein with five membrane-spanning domains and two large N-glycosylated extracellular loops that is localized to plasma membrane protrusions and microdomains (reviewed by Bauer et al.[35]). The function of prominin-1 is not entirely known, however. A single nucleotide deletion of *PROMININ-1* is responsible for an inherited form of retinal degeneration.[36] Despite the unknown cellular function of prominin-1, it has been found to be a marker for many of the CSCs identified to date, including those from gliomas,[37,38] colon,[39,40] lung,[41] liver,[42] and prostate[43] (Table 1–1). Although prominin-1 marks a tumor-initiating population in many solid tumors, it does not appear to have a significant role in maintaining properties of CSCs. In colorectal tumor cells isolated from patients, the knockdown of prominin-1 did not result in any significant decrease in the tumorigenic capacity of these cells.[44] However, the knockdown of CD44 inhibited tumor formation in these same cells.[44]

CD44 is a glycoprotein that is the receptor for hyaluronan (HA), a major component of the extracellular matrix (reviewed by Misra et al.[45]). As a result of binding HA, CD44 activates many receptor tyrosine kinases, including EGFR and ERBB2, in many cancer types.[46] This leads to increased proliferation and survival via activation of the MAPK and PI3K/AKT pathways, respectively.[47] CD44 also plays an important role in invasion of a variety of tumor cells, including breast,[48] prostate,[49] and mesotheliomas,[50] and in lymphocyte homing to the bone marrow[51] and has been positively correlated with the number of circulating prostate cancer cells in the bloodstream.[52] CD44, either alone or in combination with other surface markers, has been used to isolate cells with stem cell properties from multiple tumor types, including breast,[12,53] prostate,[54,55] colon,[56] pancreas,[57] and head and neck squamous cell carcinomas[58] (Table 1–1).

Table 1–1. Cell surface phenotypes of cancer stem cells

Tumor type	Phenotype	Fraction (%)	Reference
Breast	CD44$^+$CD24$^-$	11–35	(12)
Brain	CD133$^+$	5–30	(37, 38)
Prostate	CD44$^+$CD133$^+\alpha2\beta1^{hi}$ or	0.1–3	(43)
	CD44$^+$CD24$^-$		(55)
Pancreatic	CD44$^+$CD24$^-$ESA$^+$	0.2–0.8	(57)
Hepatocellular	CD133$^+$	1–3	(42)
Colon	CD133$^+$or ESA^{hi}CD44$^+$	1.8–24.5	(39, 40, 56)
Head and neck	CD44$^+$	<10	(58)
Lung	CD133$^+$	0.3–22	(41)
AML	CD34$^+$CD38$^+$	0.2–1	(10, 60)
Multiple myeloma	CD138$^+$	2–5	(59)
Melanoma	CD20$^+$	~20	(65)

Other markers used for the identification of CSCs tend to be more specific to the organ, and the choice is generally gleaned from knowledge of how that tissue develops. For instance, CD138 is a marker for terminally differentiated B cells (plasma), and multiple myeloma (a plasma cell malignancy) CSCs are CD138$^-$ cells.[59] Likewise, CSCs from acute myelogenous leukemia are CD34$^+$CD38$^-$ cells,[10,60] the same markers used to identify normal early hematopoietic progenitor cells.

Although using surface markers allows for the definition of a precise population, as opposed to both SPs and spheres, there are several limitations to this method of isolation as well. The number of CSCs usually identified by this method is almost always low (Table 1–1), requiring a large number of cells to be sorted. This is especially problematic when isolating cells from tumor samples that are often small in size. Also, isolation of CSCs from tissue requires the cells to be enzymatically dissociated, usually with collagenase and other proteolytic enzymes, which can damage some of the surface antigens expressed.[61] Probably the greatest drawback of using surface markers for the identification of CSCs is the choice of the markers themselves. As outlined earlier, the markers often come from what is known about the development of the tissue and from markers of stem cells from systems in which a hierarchy of differentiation has been clearly established such as CD133 from the hematopoietic system.

Culture of nonadherent spheres

In addition to both SPs and cell surface markers, CSCs have been isolated by their ability to form spheres in culture. The ability of CSCs to form spheres in culture was first demonstrated in the central nervous system. In 1992, Reynolds and Weiss demonstrated that cells isolated from the striatum of adult mouse brain could be clonally expanded by culturing spheres and that these cells could generate both astrocytes and neurons.[62] In humans, CD133$^+$ cells isolated from human fetal brain were shown to form spheres in vitro.[37] Further studies have demonstrated that brain tumors also contain CD133[+] cells that are capable of giving rise to neurospheres.[37] Subsequently, the ability of purified CSCs to form spheres in culture has been demonstrated for breast,[12,53] prostate,[54,55] colon,[40,63] pancreatic,[64] and melanoma CSCs.[65]

Because it has been demonstrated that purified CSCs can give rise to spheres in culture, some researchers have used sphere cultures to enrich CSCs. For example, it was shown that cultures of breast cancer under low-adherent sphere-forming conditions enriched the CD44[+]CD24[−] population, the surface marker phenotype associated with breast CSCs, by 40% to 98% and that the spheres were more tumorigenic in immunocompromised mice.[66] Likewise, researchers have enriched for CSCs from brain,[37,38,67,68] colon,[40] pancreas,[64] bone sarcomas,[69] and melanomas[65] by using sphere culture conditions. In all cases, these spheres are enriched in the surface markers reported by others to represent the tumor-initiating population in the respective tissue, except for the spheres generated from bone sarcomas, in which surface marker expression remains to be determined.

Although culturing for spheres is an easier method of enrichment in comparison to sorting for SPs or surface markers, it is not without limitations. Perhaps the biggest drawback is that the spheres still represent a heterogeneous population, with only a portion of the cells capable of self-renewal.[66,70] Furthermore, immunohistochemical staining of spheres generated from prostate cell lines show that the spheres are heterogeneous for markers of CSCs.[54] Furthermore, differences in the enrichment of CSCs in spheres due to differences in sphere size, passage, culture medium, and technique can be demonstrated in neurosphere cultures.[71]

PROPERTIES AND CHARACTERIZATION OF CSCs

Despite the isolation methodology, establishing that a subpopulation of cells is indeed a CSC population relies on validation of several of the biological characteristics of CSCs, including tumorigenicity, self-renewal, and the ability to histologically recapitulate the tumor of origin. Indeed, a 2006 American Association for Cancer Research workshop concluded that "cancer stem cells can thus, only be defined experimentally by their ability to recapitulate the generation of a continuously growing tumor."[72]

Tumorigenicity

At the heart of the definition of CSCs is their ability to induce tumor formation. Most experiments demonstrating tumorigenicity utilize one of two immunodeficient mouse models: the nude mouse or the Non-obese diabetic/severe combined immunodeficient (NOD/SCID) mouse. The nude mouse, a result of a mutation in the *FOXN* gene, is athymic, resulting in the hairless phenotype that gives it its name and in a lack of mature T cells.[73] The NOD/SCID mouse model[74]

Table 1–2. Numbers of CSCs required for tumor formation and sites of injection

Tumor type	Site of injection	Numbers of cells tested	Lowest number required	Reference
Brain	Brain	100–100,000	100	(38)
Prostate	Subcutaneous	100–10,000	100[a]	(54, 55)
	Prostate	100–1,000,000	1,000	(79)
Pancreatic	Subcutaneous	100–10,000	100	(57)
Hepatocellular	Intrahepatic	50,000–300,000	50,000	(42)
	Subcutaneous	1,000–1,000,000	1,000	(42)
Colon	Renal capsule	100–250,000	100	(39)
	Subcutaneous	3,000–100,000	3,000	(40)
Head and neck	Subcutaneous	2,000–650,000	5,000	(58)
Lung	Subcutaneous	10,000–500,000	10,000	(41)

[a] Ten CD44+ cells from a single cell line, LAPC-9, were able to induce a tumor in one-fourth of mice tested. This was the only cell line in which 10 cells were tested.

is the result of a cross of the SCID mouse model, which lacks both T and B lymphocytes,[75] and the NOD mouse model, which lacks natural killer cells and antigen-presenting cells. The result is a mouse model that has functional defects in both innate and adaptive immunity. Both models result in a mouse that does not reject xenografts.

Historically, researchers needed to inject millions of cells to establish a tumor. This was first demonstrated in 1961, when researchers harvested recurrent cancer cells from patients and then autotransplanted the cells. Tumors formed only when patients received injections of one million cells.[7] The requirement for millions of cells to establish a tumor led investigators to hypothesize that there are only a limited number of cells that are able to initiate and maintain the tumor. Theoretically, implantation of a single CSC should be capable of generating the entire tumor in a mouse model. Therefore one important test of a prospective CSC population is the ability to form tumors at low cell densities. Most tests of CSC-induced tumorigenesis have used anywhere between 100 and 1,000 cells as the lowest number of cells injected (Table 1–2). In four cancers – brain,[38] prostate,[54,55] colon,[39] and pancreatic[57] – as few as 100 cells were able to give rise to tumor formation in a significant number of the animals tested. Patrawala and colleagues were able to show that 10 CD44+ cells isolated from the prostate cancer cell line LAPC-9 were able to give rise to a tumor in one-fourth of mice.[54]

Self-renewal

The CSC must have the ability to sustain itself and continue to give rise to cells with equal abilities of tumorigenicity and recapitulation of the original tumor. This ability of the CSC to give rise to another CSC is termed *self-renewal*. Self-renewal maintains a reservoir of CSCs when the CSCs undergoes either asymmetrical or symmetrical division (reviewed by Huntly and Gilliland[6]). Asymmetric division forms one more differentiated cell and one CSC. Symmetrical division results in the CSC forming either two differentiated cells or two CSCs. This behavior is critical because it allows the CSC to expand its numbers.

Self-renewal has been experimentally demonstrated in two major ways: (1) by serial transplantation of tumors and (2) by showing the ability of CSCs to initiate spheres or soft agar colonies over multiple generations. Although serial transplantation of a tumor is the most rigorous proof of the ability of the CSC to self-renew, it is also lengthy and more expensive than culture techniques. Culture methods rely on the assumption that the ability to form a sphere or a colony in soft agar is a surrogate for tumor formation. As long as the spheres or colonies have been shown to be more tumorigenic than nonspheres or the total cell population in a mouse model, this is a fair assumption.

Serial transplantation of CSCs

Serial transplantation involves isolating the prospective CSCs, initiating a tumor in a mouse model, and subsequently removing the tumor to reisolate the cells with the prospective CSC phenotype and retesting tumorigenicity with these isolated CSCs. In theory, the CSCs isolated from any generation of tumor should be able to give rise to a subsequent tumor. The first isolation of CSCs from a solid tumor was from breast cancer by al-Hajj and colleagues in 2003.[12] They found that as few as 100 $CD44^+CD24^-$ cells could form a tumor in immunocompromised NOD/SCID mice. Furthermore, $CD44^+CD24^-$ cells were able to give rise to tumors when serially transplanted into NOD/SCID mice through four passages. The vast majority of cells isolated within the primary and subsequent tumors were more differentiated, and these cells were unable to generate a tumor. This provided the most compelling demonstration that the $CD44^+CD24^-$ breast cancer cells were indeed CSCs. Since this initial demonstration of self-renewal by serial transplantation, CSCs isolated from several solid tissues have been demonstrated to have self-renewal capabilities by serial transplantation. These include CSCs isolated from brain (two generations),[38] prostate (two generations),[54] pancreas (two generations),[57] hepatocellular carcinoma (two generations),[42] colon carcinoma (three and four generations, respectively),[39,40] and head and neck carcinomas (three generations).[58]

In vitro renewal of CSCs

Demonstration of the ability to self-renew in a culture system provides a shorter, less expensive alternative to mouse models. This in vitro technique employs the same principal as the in vivo self-renewal assay. Nonadherent spheres or colonies in soft agar are formed, dissociated, and replanted to determine the ability of the cells to form new spheres or colonies. Several CSCs have been shown to have in vitro self-renewal capacities, as measured by their ability to form spheres or colonies through multiple generations. For example, Ricci-Vitiani and colleagues were able to demonstrate the ability of spheres derived from the colon to reform spheres up to 10 generations.[40] They also demonstrated that the spheres are able to induce tumor formation. Similar in vitro self-renewal assays have been performed with CSCs isolated from prostate cancer[76] and lung cancer.[41]

Establishment of tumor heterogeneity

Not only must a CSC be able to self-renew, but it must also be able to differentiate to recapitulate the heterogeneity seen in tumors (reviewed by al-Hajj and Clarke[77]). They are the putative population responsible for generation and maintenance of a heterogeneous population of cells. Again, two approaches have been taken: one is to examine the heterogeneity of CSC-derived tumors, and the other is to determine the ability of CSCs to differentiate in vitro.

The heterogeneity of a CSC-derived tumor has been demonstrated either by flow cytometry of surface markers or by immunohistochemistry. Al-Hajj and colleagues were the first to demonstrate the ability of a CSC to give rise to a heterogeneous tumor population.[12] They showed, by flow cytometry, that injection of mammary CD44+CD24− cells led to tumors with a diverse surface phenotype reminiscent of the original tumor, with only a minority of cells retaining the CD44+CD24− phenotype. Since then, all solid-tumor CSCs listed in Table 1–1 have been shown to differentiate into other cell types.

The ability of a CSC to differentiate in vitro into the other cell types present in a tumor has also been demonstrated. This has been demonstrated for brain,[78] prostate,[43,55] colon,[40] and lung cancers.[41] Differentiation in culture has been shown to occur in the presence of serum both with and without other factors known to induce differentiation in the specific tissue type being studied. The in vitro differentiated cells not only lose markers of CSCs, but also lose tumorigenic potential.[41,78]

CONCLUDING REMARKS

The field of CSC research is a rapidly moving field that is still in its infancy. The information that we glean from CSCs isolated from one type of cancer is not always applicable to another cancer, type. This has already been the case when choosing cell surface markers for the isolation of the CSCs. It is too early to take much of what has been hypothesized and determined for a single cancer and use it as dogma. Therefore, before further classifying a subpopulation of cells, no matter the method of isolation, investigators must first determine if the population has properties of CSCs, including self-renewal, increased tumorigenic potential, and the ability to recapitulate tumor heterogeneity.

ACKNOWLEDGMENTS

This publication has been funded in part with federal funds from the National Cancer Institute, National Institutes of Health (NIH), under contract N01-CO-12400. This research was supported in part by the Intramural Research Program of the NIH, National Cancer Institute.

REFERENCES

1. Bignold, L.P., Coghlan, B.L., and Jersmann, H.P. Hansemann, Boveri, chromosomes and the gametogenesis-related theories of tumours, Cell Biol. Int., *30:* 640–644, 2006.
2. Sell, S. Stem cell origin of cancer and differentiation therapy, Crit. Rev. Oncol. Hematol., *51:* 1–28, 2004.
3. Knudson, A.G., Jr., Strong, L.C., and Anderson, D.E. Heredity and cancer in man, Prog. Med. Genet., *9:* 113–158, 1973.
4. Wang, J.C., and Dick, J.E. Cancer stem cells: lessons from leukemia, Trends Cell Biol., *15:* 494–501, 2005.
5. Reya, T., Morrison, S.J., Clarke, M.F., and Weissman, I.L. Stem cells, cancer, and cancer stem cells, Nature, *414:* 105–111, 2001.
6. Huntly, B.J., and Gilliland, D.G. Leukaemia stem cells and the evolution of cancer-stem-cell research, Nat. Rev. Cancer, *5:* 311–321, 2005.
7. Southam, C.M., and Brunschwig, A. Quantitative studies of autotransplantation of human cancer, Cancer, *14:* 971–978, 1961.
8. Bruce, W.R., and Van Der Gaag, H. ADE quantitative assay for the number of murine lymphoma cells capable of proliferation in vivo, Nature, *199:* 79–80, 1963.
9. Hamburger, A.W., and Salmon, S.E. Primary bioassay of human tumor stem cells, Science, *197:* 461–463, 1977.
10. Lapidot, T., Sirard, C., Vormoor, J., Murdoch, B., Hoang, T., Caceres-Cortes, J., Minden, M., Paterson, B., Caligiuri, M.A., and Dick, J.E. A cell initiating human acute myeloid leukaemia after transplantation into SCID mice, Nature, *367:* 645–648, 1994.
11. Al-Hajj, M., Wicha, M.S., ito-Hernandez, A., Morrison, S.J., and Clarke, M.F. Prospective identification of tumorigenic breast cancer cells, Proc. Natl. Acad. Sci. U. S. A., *100:* 3983–3988, 2003.
12. Goodell, M.A., Brose, K., Paradis, G., Conner, A.S., and Mulligan, R.C. Isolation and functional properties of murine hematopoietic stem cells that are replicating in vivo, J. Exp. Med., *183:* 1797–1806, 1996.
13. Wulf, G.G., Wang, R.Y., Kuehnle, I., Weidner, D., Marini, F., Brenner, M.K., Andreeff, M., and Goodell, M.A. A leukemic stem cell with intrinsic drug efflux capacity in acute myeloid leukemia, Blood, *98:* 1166–1173, 2001.
14. Feuring-Buske, M., and Hogge, D.E. Hoechst 33342 efflux identifies a subpopulation of cytogenetically normal $CD34^+CD38^-$ progenitor cells from patients with acute myeloid leukemia, Blood, *97:* 3882–3889, 2001.
15. Szotek, P.P., Pieretti-Vanmarcke, R., Masiakos, P.T., Dinulescu, D.M., Connolly, D., Foster, R., Dombkowski, D., Preffer, F., Maclaughlin, D.T., and Donahoe, P.K. Ovarian cancer side population defines cells with stem cell-like characteristics and Mullerian Inhibiting Substance responsiveness, Proc. Natl. Acad. Sci. U. S. A., *103:* 11154–11159, 2006.
16. Chiba, T., Kita, K., Zheng, Y.W., Yokosuka, O., Saisho, H., Iwama, A., Nakauchi, H., and Taniguchi, H. Side population purified from hepatocellular carcinoma cells harbors cancer stem cell-like properties, Hepatology, *44:* 240–251, 2006.
17. Hirschmann-Jax, C., Foster, A.E., Wulf, G.G., Nuchtern, J.G., Jax, T.W., Gobel, U., Goodell, M.A., and Brenner, M.K. A distinct "side population" of cells with high drug efflux capacity in human tumor cells, Proc. Natl. Acad. Sci. U. S. A., *101:* 14228–14233, 2004.
18. Setoguchi, T., Taga, T., and Kondo, T. Cancer stem cells persist in many cancer cell lines, Cell Cycle, *3:* 414–415, 2004.
19. Shen, G., Shen, F., Shi, Z., Liu, W., Hu, W., Zheng, X., Wen, L., and Yang, X. Identification of cancer stem-like cells in the C6 glioma cell line and the limitation of current identification methods, In Vitro Cell Dev. Biol. Anim., *44:* 280–289, 2008.
20. Ho, M.M., Ng, A.V., Lam, S., and Hung, J.Y. Side population in human lung cancer cell lines and tumors is enriched with stem-like cancer cells, Cancer Res., *67:* 4827–4833, 2007.

21. Mitsutake, N., Iwao, A., Nagai, K., Namba, H., Ohtsuru, A., Saenko, V., and Yamashita, S. Characterization of side population in thyroid cancer cell lines: cancer stem-like cells are enriched partly but not exclusively, Endocrinology, *148:* 1797–1803, 2007.

22. Wang, J., Guo, L.P., Chen, L.Z., Zeng, Y.X., and Lu, S.H. Identification of cancer stem cell-like side population cells in human nasopharyngeal carcinoma cell line, Cancer Res., *67:* 3716–3724, 2007.

23. Patrawala, L., Calhoun, T., Schneider-Broussard, R., Zhou, J., Claypool, K., and Tang, D.G. Side population is enriched in tumorigenic, stem-like cancer cells, whereas ABCG2$^+$ and ABCG2$^-$ cancer cells are similarly tumorigenic, Cancer Res., *65:* 6207–6219, 2005.

24. Goodell, M.A., Kinney-Freeman, S., and Camargo, F.D. Isolation and characterization of side population cells, Methods Mol. Biol., *290:* 343–352, 2005.

25. Zhou, S., Morris, J.J., Barnes, Y., Lan, L., Schuetz, J.D., and Sorrentino, B.P. Bcrp1 gene expression is required for normal numbers of side population stem cells in mice, and confers relative protection to mitoxantrone in hematopoietic cells in vivo, Proc. Natl. Acad. Sci. U. S. A., *99:* 12339–12344, 2002.

26. Yin, L., Castagnino, P., and Assoian, R.K. ABCG2 expression and side population abundance regulated by a transforming growth factor beta-directed epithelial-mesenchymal transition, Cancer Res., *68:* 800–807, 2008.

27. Bhatt, R.I., Brown, M.D., Hart, C.A., Gilmore, P., Ramani, V.A., George, N.J., and Clarke, N.W. Novel method for the isolation and characterisation of the putative prostatic stem cell, Cytometry A, *54:* 89–99, 2003.

28. Smalley, M.J., and Clarke, R.B. The mammary gland "side population": a putative stem/progenitor cell marker?, J. Mammary Gland Biol. Neoplasia, *10:* 37–47, 2005.

29. Iwatani, H., Ito, T., Imai, E., Matsuzaki, Y., Suzuki, A., Yamato, M., Okabe, M., and Hori, M. Hematopoietic and nonhematopoietic potentials of Hoechst(low)/side population cells isolated from adult rat kidney, Kidney Int., *65:* 1604–1614, 2004.

30. Terunuma, A., Jackson, K.L., Kapoor, V., Telford, W.G., and Vogel, J.C. Side population keratinocytes resembling bone marrow side population stem cells are distinct from label-retaining keratinocyte stem cells, J. Invest. Dermatol., *121:* 1095–1103, 2003.

31. Triel, C., Vestergaard, M.E., Bolund, L., Jensen, T.G., and Jensen, U.B. Side population cells in human and mouse epidermis lack stem cell characteristics, Exp. Cell Res., *295:* 79–90, 2004.

32. Weigmann, A., Corbeil, D., Hellwig, A., and Huttner, W.B. Prominin, a novel microvilli-specific polytopic membrane protein of the apical surface of epithelial cells, is targeted to plasmalemmal protrusions of non-epithelial cells, Proc. Natl. Acad. Sci. U. S. A., *94:* 12425–12430, 1997.

33. Yin, A.H., Miraglia, S., Zanjani, E.D., Almeida-Porada, G., Ogawa, M., Leary, A.G., Olweus, J., Kearney, J., and Buck, D.W. AC133, a novel marker for human hematopoietic stem and progenitor cells, Blood, *90:* 5002–5012, 1997.

34. Bauer, N., Fonseca, A.V., Florek, M., Freund, D., Jaszai, J., Bornhauser, M., Fargeas, C.A., and Corbeil, D. New insights into the cell biology of hematopoietic progenitors by studying prominin-1 (CD133), Cells Tissues Organs, *188:* 127–138, 2008.

35. Maw, M.A., Corbeil, D., Koch, J., Hellwig, A., Wilson-Wheeler, J.C., Bridges, R.J., Kumaramanickavel, G., John, S., Nancarrow, D., Roper, K., Weigmann, A., Huttner, W.B., and Denton, M.J. A frameshift mutation in prominin (mouse)-like 1 causes human retinal degeneration, Hum. Mol. Genet., *9:* 27–34, 2000.

36. Singh, S.K., Clarke, I.D., Terasaki, M., Bonn, V.E., Hawkins, C., Squire, J., and Dirks, P.B. Identification of a cancer stem cell in human brain tumors, Cancer Res., *63:* 5821–5828, 2003.

37. Singh, S.K., Hawkins, C., Clarke, I.D., Squire, J.A., Bayani, J., Hide, T., Henkelman, R.M., Cusimano, M.D., and Dirks, P.B. Identification of human brain tumour initiating cells, Nature, *432:* 396–401, 2004.

38. O'Brien, C.A., Pollett, A., Gallinger, S., and Dick, J.E. A human colon cancer cell capable of initiating tumour growth in immunodeficient mice, Nature, *445:* 106–110, 2007.

39. Ricci-Vitiani, L., Lombardi, D.G., Pilozzi, E., Biffoni, M., Todaro, M., Peschle, C., and De, Maria, R. Identification and expansion of human colon-cancer-initiating cells, Nature, *445:* 111–115, 2007.

40. Eramo, A., Lotti, F., Sette, G., Pilozzi, E., Biffoni, M., Di, Virgilio, A., Conticello, C., Ruco, L., Peschle, C., and De, Maria, R. Identification and expansion of the tumorigenic lung cancer stem cell population, Cell Death Differ., *15:* 504–514, 2008.

41. Ma, S., Chan, K.W., Hu, L., Lee, T.K., Wo, J.Y., Ng, I.O., Zheng, B.J., and Guan, X.Y. Identification and characterization of tumorigenic liver cancer stem/progenitor cells, Gastroenterology, *132:* 2542–2556, 2007.

42. Collins, A.T., Berry, P.A., Hyde, C., Stower, M.J., and Maitland, N.J. Prospective identification of tumorigenic prostate cancer stem cells, Cancer Res., *65:* 10946–10951, 2005.

43. Du, L., Wang, H., He, L., Zhang, J., Ni, B., Wang, X., Jin, H., Cahuzac, N., Mehrpour, M., Lu, Y., and Chen, Q. CD44 is of functional importance for colorectal cancer stem cells, Clin. Cancer Res., *14:* 6751–6760, 2008.

44. Misra, S., Hascall, V.C., Berger, F.G., Markwald, R.R., and Ghatak, S. Hyaluronan, CD44, and cyclooxygenase-2 in colon cancer, Connect. Tissue Res., *49:* 219–224, 2008.

45. Ghatak, S., Misra, S., and Toole, B.P. Hyaluronan constitutively regulates ErbB2 phosphorylation and signaling complex formation in carcinoma cells, J. Biol. Chem., *280:* 8875–8883, 2005.

46. Misra, S., Toole, B.P., and Ghatak, S. Hyaluronan constitutively regulates activation of multiple receptor tyrosine kinases in epithelial and carcinoma cells, J. Biol. Chem., *281:* 34936–34941, 2006.

47. Sheridan, C., Kishimoto, H., Fuchs, R.K., Mehrotra, S., Bhat-Nakshatri, P., Turner, C.H., Goulet, R., Jr., Badve, S., and Nakshatri, H. CD44+/CD24− breast cancer cells exhibit enhanced invasive properties: an early step necessary for metastasis, Breast Cancer Res., *8:* R59, 2006.

48. Omara-Opyene, A.L., Qiu, J., Shah, G.V., and Iczkowski, K.A. Prostate cancer invasion is influenced more by expression of a CD44 isoform including variant 9 than by Muc18, Lab. Invest., *84:* 894–907, 2004.

49. Li, Y., and Heldin, P. Hyaluronan production increases the malignant properties of mesothelioma cells, Br. J. Cancer, *85:* 600–607, 2001.

50. Avigdor, A., Goichberg, P., Shivtiel, S., Dar, A., Peled, A., Samira, S., Kollet, O., Hershkoviz, R., Alon, R., Hardan, I., Ben-Hur, H., Naor, D., Nagler, A., and Lapidot, T. CD44 and hyaluronic acid cooperate with SDF-1 in the trafficking of human CD34+ stem/progenitor cells to bone marrow, Blood, *103:* 2981–2989, 2004.

51. Paradis, V., Eschwege, P., Loric, S., Dumas, F., Ba, N., Benoit, G., Jardin, A., and Bedossa, P. De novo expression of CD44 in prostate carcinoma is correlated with systemic dissemination of prostate cancer, J. Clin. Pathol., *51:* 798–802, 1998.

52. Dontu, G., Abdallah, W.M., Foley, J.M., Jackson, K.W., Clarke, M.F., Kawamura, M.J., and Wicha, M.S. In vitro propagation and transcriptional profiling of human mammary stem/progenitor cells, Genes Dev., *17:* 1253–1270, 2003.

53. Patrawala, L., Calhoun, T., Schneider-Broussard, R., Li, H., Bhatia, B., Tang, S., Reilly, J.G., Chandra, D., Zhou, J., Claypool, K., Coghlan, L., and Tang, D.G. Highly purified CD44+ prostate cancer cells from xenograft human tumors are enriched in tumorigenic and metastatic progenitor cells, Oncogene, *25:* 1696–1708, 2006.

54. Hurt, E.M., Kawasaki, B.T., Klarmann, G.J., Thomas, S.B., and Farrar, W.L. CD44(+)CD24(−) prostate cells are early cancer progenitor/stem cells that provide a model for patients with poor prognosis, Br. J. Cancer, *98:* 756–765, 2008.

55. Dalerba, P., Dylla, S.J., Park, I.K., Liu, R., Wang, X., Cho, R.W., Hoey, T., Gurney, A., Huang, E.H., Simeone, D.M., Shelton, A.A., Parmiani, G., Castelli, C., and Clarke, M.F. Phenotypic characterization of human colorectal cancer stem cells, Proc. Natl. Acad. Sci. U. S. A., *104:* 10158–10163, 2007.

56. Li, C., Heidt, D.G., Dalerba, P., Burant, C.F., Zhang, L., Adsay, V., Wicha, M., Clarke, M.F., and Simeone, D.M. Identification of pancreatic cancer stem cells, Cancer Res., *67:* 1030–1037, 2007.

57. Prince, M.E., Sivanandan, R., Kaczorowski, A., Wolf, G.T., Kaplan, M.J., Dalerba, P., Weissman, I.L., Clarke, M.F., and Ailles, L.E. Identification of a subpopulation of cells with cancer stem cell properties in head and neck squamous cell carcinoma, Proc. Natl. Acad. Sci. U. S. A, *104:* 973–978, 2007.
58. Matsui, W., Huff, C.A., Wang, Q., Malehorn, M.T., Barber, J., Tanhehco, Y., Smith, B.D., Civin, C.I., and Jones, R.J. Characterization of clonogenic multiple myeloma cells, Blood, *103:* 2332–2336, 2004.
59. Bonnet, D., and Dick, J.E. Human acute myeloid leukemia is organized as a hierarchy that originates from a primitive hematopoietic cell, Nat. Med., *3:* 730–737, 1997.
60. Masters, J.R., Foley, C.L., Bisson, I., and Ahmed, A. Cancer stem cells, BJU Int., *92:* 661–662, 2003.
61. Reynolds, B.A., and Weiss, S. Generation of neurons and astrocytes from isolated cells of the adult mammalian central nervous system, Science, *255:* 1707–1710, 1992.
62. Todaro, M., Alea, M.P., Di Stefano, A.B., Cammareri, P., Vermeulen, L., Iovino, F., Tripodo, C., Russo, A., Gulotta, G., Medema, J.P., and Stassi, G. Colon cancer stem cells dictate tumor growth and resist cell death by production of interleukin-4, Cell Stem Cell, *1:* 389–402, 2007.
63. Gou, S., Liu, T., Wang, C., Yin, T., Li, K., Yang, M., and Zhou, J. Establishment of clonal colony-forming assay for propagation of pancreatic cancer cells with stem cell properties, Pancreas, *34:* 429–435, 2007.
64. Fang, D., Nguyen, T.K., Leishear, K., Finko, R., Kulp, A.N., Hotz, S., Van Belle, P.A., Xu, X., Elder, D.E., and Herlyn, M. A tumorigenic subpopulation with stem cell properties in melanomas, Cancer Res., *65:* 9328–9337, 2005.
65. Ponti, D., Costa, A., Zaffaroni, N., Pratesi, G., Petrangolini, G., Coradini, D., Pilotti, S., Pierotti, M.A., and Daidone, M.G. Isolation and in vitro propagation of tumorigenic breast cancer cells with stem/progenitor cell properties, Cancer Res., *65:* 5506–5511, 2005.
66. Yuan, X., Curtin, J., Xiong, Y., Liu, G., Waschsmann-Hogiu, S., Farkas, D.L., Black, K.L., and Yu, J.S. Isolation of cancer stem cells from adult glioblastoma multiforme, Oncogene, *23:* 9392–9400, 2004.
67. Galli, R., Binda, E., Orfanelli, U., Cipelletti, B., Gritti, A., De Vitis, S., Fiocco, R., Foroni, C., Dimeco, F., and Vescovi, A. Isolation and characterization of tumorigenic, stem-like neural precursors from human glioblastoma, Cancer Res., *64:* 7011–7021, 2004.
68. Gibbs, C.P., Kukekov, V.G., Reith, J.D., Tchigrinova, O., Suslov, O.N., Scott, E.W., Ghivizzani, S.C., Ignatova, T.N., and Steindler, D.A. Stem-like cells in bone sarcomas: implications for tumorigenesis, Neoplasia, *7:* 967–976, 2005.
69. Suslov, O.N., Kukekov, V.G., Ignatova, T.N., and Steindler, D.A. Neural stem cell heterogeneity demonstrated by molecular phenotyping of clonal neurospheres, Proc. Natl. Acad. Sci. U. S. A, *99:* 14506–14511, 2002.
70. Jensen, J.B., and Parmar, M. Strengths and limitations of the neurosphere culture system, Mol. Neurobiol., *34:* 153–161, 2006.
71. Clarke, M.F., Dick, J.E., Dirks, P.B., Eaves, C.J., Jamieson, C.H., Jones, D.L., Visvader, J., Weissman, I.L., and Wahl, G.M. Cancer stem cells – perspectives on current status and future directions: AACR Workshop on Cancer Stem Cells, Cancer Res., *66:* 9339–9344, 2006. Quote from p. 9340.
72. Flanagan, S.P. "Nude," a new hairless gene with pleiotropic effects in the mouse, Genet. Res., *8:* 295–309, 1966.
73. Shultz, L.D., Schweitzer, P.A., Christianson, S.W., Gott, B., Schweitzer, I.B., Tennent, B., McKenna, S., Mobraaten, L., Rajan, T.V., and Greiner, D.L. Multiple defects in innate and adaptive immunologic function in NOD/LtSz-SCID mice, J. Immunol., *154:* 180–191, 1995.
74. Bosma, G.C., Custer, R.P., and Bosma, M.J. A severe combined immunodeficiency mutation in the mouse, Nature, *301:* 527–530, 1983.
75. Collins, A.T., Habib, F.K., Maitland, N.J., and Neal, D.E. Identification and isolation of human prostate epithelial stem cells based on alpha(2)beta(1)-integrin expression, J. Cell Sci., *114:* 3865–3872, 2001.

76. Al-Hajj, M., and Clarke, M.F. Self-renewal and solid tumor stem cells, Oncogene, *23:* 7274–7282, 2004.

77. Lee, J., Kotliarova, S., Kotliarov, Y., Li, A., Su, Q., Donin, N.M., Pastorino, S., Purow, B.W., Christopher, N., Zhang, W., Park, J.K., and Fine, H.A. Tumor stem cells derived from glioblastomas cultured in bFGF and EGF more closely mirror the phenotype and genotype of primary tumors than do serum-cultured cell lines, Cancer Cell, *9:* 391–403, 2006.

78. Tang, D.G., Patrawala, L., Calhoun, T., Bhatia, B., Choy, G., Schneider-Broussard, R., and Jeter, C. Prostate cancer stem/progenitor cells: identification, characterization, and implications, Mol. Carcinog., *46:* 1–14, 2007.

2 Prostate cancer stem cells

Collene R. Jeter and Dean G. Tang

University of Texas M. D. Anderson Cancer Center

STEM CELLS, PROGENITOR CELLS, AND DIFFERENTIATED CELLS

Functional regeneration, the ability of cells to reconstitute the tissue of origin, is an essential biological property of many epithelia. This unique ability suggests the presence of a renewing cell type and reflects the homeostatic mechanism that normally replaces senescent cells or cells lost to tissue damage. Not all cells within a population are equally capable of reconstitution, and this activity has been attributed to the presence of a subset of tissue-specific stem and/or progenitor cells within various epithelia, including the breast,[1-4] skin,[5-7] intestine,[8] and, of particular interest, the prostate.[9-11]

Cellular hierarchy is essential to the biology of complex multicellular organisms, and aberrant cell fate determination may result in pathological phenotypes. During embryogenesis, a phenomenal array of specialized cells arises from primitive, undifferentiated stem cells (SCs). The rapidly dividing cells of the early blastocyst inner-cell mass, and their derived cultured counterparts, termed *embryonic stem cells* (ESCs), exhibit pluripotency and unlimited proliferative potential.[12] Both extrinsic signals and intrinsic properties converge to activate precise differentiation programs, thereby generating the phenotypically and functionally distinct lineage-restricted daughter cells present in the developing fetus.

Growth and maturation require the continual activity of stem-like cells after birth. These somatic SCs also function to repair tissue damage and maintain tissue homeostasis over time.[1,5,6,13–15] All SCs possess remarkable proliferative potential; however, unlike ESCs, somatic SCs rarely divide. Somatic SC division is constrained by interactions within a specialized stromal cell and extracellular matrix–rich environment (the SC niche). Although generally quiescent, in response to stimulatory conditions, somatic SCs can reenter the cell cycle and divide.[16,17] In the classical hierarchical model, these slow-cycling SCs are usually multipotent (although this may vary widely in a tissue-dependent manner) and give rise to more rapidly dividing progenitor cells that, in turn, generate the terminally differentiated (and often nondividing) functional cells of the tissue or organ.

SC division resulting in at least one daughter cell maintaining the hallmark features of SCs is *self-renewal*, and the long-term regenerative activities of SCs depend on this essential property.[12,18,19] Asymmetric SC division also produces a more rapidly dividing lineage-restricted progenitor cell, which has lost (or reduced) self-renewal potential. In this way, the SC population is maintained while contributing to the wide variety of cell types in a given organ.[6] Altogether, this heterogeneous mixture of cells with differing proliferative capacities and differentiation states constitutes the functional epithelium, and SCs account for the variety of cell types present.

CANCER STEM CELLS (CSCs)

The elegance of metazoan tissue organization is a delicate balance, one that is disrupted during tumorigenesis. Tumor formation has been described as "ontogeny gone awry" as tumor cells, like embryonic cells during organogenesis, exhibit a high proliferative capacity and multipotency. Therefore, conceptually, tumors may contain stem-like cells or CSCs. In support, although most tumors arise via the clonal expansion of a single transformed cell, tumors are usually heterogeneous, containing multiple types of cells ranging in maturation. Although genetic instability can give rise to a mixture of cells with various mutations, a plausible alternative explanation for tumor heterogeneity is that abnormal SCs present in the tumor can generate progenitor and differentiated daughter cells by epigenetic processes. The earliest experimental evidence supporting the existence of stem-like cells within tumors came more than 40 years ago and showed that only a small percentage of tumor cells is tumorigenic in vivo or clonogenic in vitro.[18,20–22]

Cancer may be considered a stem cell disease.[14,22] The long life of the SC might predispose these cells to transformation, as the multiple mutation events required for tumor initiation can be acquired over time, or alternatively, some tumor cells may acquire SC-like properties. In particular, the acquisition of self-renewal by a more differentiated cell may impart immortalization, a cardinal requirement for tumor transformation. Functionally, CSCs are equivalent to tumor-initiating (or -reinitiating) and tumor-maintaining cells. CSCs have also been proposed to underlie tumor recurrence following therapeutic treatments. Presumably, these treatments would preferentially target the rapidly dividing progenitor cells, leaving behind drug-resistant and indolent CSCs, potentially regenerating an even more aggressive cancer.

The most definitive evidence supporting the CSC hypothesis has come from transplantation experiments, in which only a minor subset of tumor cells is tumorigenic, and these tumor-initiating (or -reinitiating) cells often display key similarities to normal SCs. With the availability of various markers that allow the prospective identification and purification of normal tissue SCs, putative CSCs were first isolated from human acute myeloid leukemia using markers that identify normal hematopoietic SCs. Remarkably, although the leukemic SCs constituted <1% of the tumor cells, they were the only cells that could transplant leukemia to nonobese diabetic/severe combined immunodeficiency disease (NOD/SCID) mice.[23] Using similar strategies (i.e., marker analysis), tumor-initiating, putative CSCs have recently been identified in solid tumors such as those derived from human breast,[24] brain,[25] colon,[26,27] and, as we shall describe in further detail, prostate.[28,29]

Phenotypic assays have also been utilized to identify (and characterize) candidate CSCs or tumor progenitors. The side population (SP) phenotype permits the discrimination of SCs (both normal and tumor-derived) from the bulk population on the basis that cells with SC properties do not accumulate Hoechst dye 33342 due to over expression of detoxifying multidrug-resistance (MDR) transporters.[30–34] Another alternative is sphere-forming assays, the ability of a cell to survive anchorage deprivation and proliferate sufficiently to give rise to a three-dimensional ball (i.e., sphere) of cells.[29,30,32]

The somatic SC property of quiescence may be another method to isolate and characterize tumorigenic stem-like cells. Slow-cycling cells can be identified as long-term label-retaining cells (LRCs) using a pulse-chase approach, in which cells are first labeled with DNA analogs, such as ^{3}H-thymidine or 5′-bromo-deoxyuridine (BrdU), followed by an extended period of chase, in which the exogenous label is diluted out via DNA replication during cell division. This LRC phenotype has been successfully used to identify normal somatic SCs from a variety of tissues, including the skin,[5,13] breast,[1] colon,[8] and prostate.[9] Although some evidence suggests that slow-cycling cells are present in tumor cell populations,[3,29,35] further work is required to determine whether these slow-cycling tumor cells are truly CSCs.

In summary, CSCs appear to be quite similar to their normal SC counterparts, especially with respect to their unique capacity to generate heterogeneous progeny. Specifically, these stem-like cells are relatively rare (i.e., representing a certain fraction of the bulk tumor cells), possess a high proliferative capacity (e.g., exhibit a high cloning efficiency in vitro), and have the ability to give rise to differentiated progeny (i.e., multipotency). Therefore bona fide CSCs must also possess the hallmark SC feature of self-renewal, thus imparting the ability to maintain the population of stem-like cells coincident with the generation of more differentiated daughter cells.

Consequently, an understanding of normal somatic SCs, including the expression of cell surface markers that can be used to distinguish hierarchical cells within the tissue as well as the molecular mechanisms regulating proliferation (or quiescence), survival, and, importantly, renewal, is critical to elucidating corresponding CSCs in tumors. With this in mind, we shall describe our current knowledge of the normal prostate and prostatic lineages, prior to a discussion of evidence pointing to the existence of prostate tumor–derived CSCs, and the possible therapeutic advantages of targeting such malignant stem-like cells in the clinic.

THE PROSTATE AND PROSTATIC STEM/PROGENITOR CELLS

The prostate is a hormonally regulated male secretory organ composed of a multitude of cells, some of which possess renewal properties.[15,36,37] Androgens stimulate prostatic growth and development during sexual maturation, and the adult prostate has a predilection to continue to grow, albeit slowly, throughout life. Although structurally distinct, both mouse and human prostate glands exhibit a tubular architecture.[38] A myriad of supportive cells, predominantly stromal cells and contractile smooth muscle cells (as well as endothelial and nerve cells), surround the acini, composed mainly of two histologically distinguishable epithelial cell types that are derived from the urogenital sinus: basal cells and secretory luminal cells[39,40] (see Figure 2–1). Basal cells are localized in a thin layer adjacent to the basement membrane surrounding each duct, with abundant exocrine luminal cells residing above proximal to the lumen. Interspersed between the layers are rare neuroendocrine cells, possibly of neural crest origin[41] or, alternatively, derived from a multipotent epithelial precursor cell.[15,42]

Complex signaling networks modulate prostate epithelial cell fate, and a key example is the action of the androgen hormone testosterone. Testosterone is enzymatically converted to a physiologically active form dihydrotestosterone (DHT), mainly in prostatic basal cells.[43] DHT then diffuses into nearby cells, where it binds to inactive androgen receptors (AR), promoting its translocation into the nucleus and the transcriptional activation of target gene expression such as differentiation markers in luminal cells (e.g., prostate specific antigen [PSA] and the homeobox transcription factor NKX3.1) and peptide growth factors in stromal cells. Transit-amplifying cell (i.e., progenitor cell) proliferative activity and luminal cell survival depend on these stromal cell–derived androgen-induced peptide growth factors.[43] As a consequence of these cell-signaling interactions, androgen deprivation in the adult (by castration or hormonal ablation) results in involution of the gland, largely due to apoptosis of the androgen-dependent luminal cells. Androgen restoration regenerates the gland, a process that can be repeated numerous times, and this renewal capacity is among the first evidence pointing to the existence of prostatic SCs.[9,44]

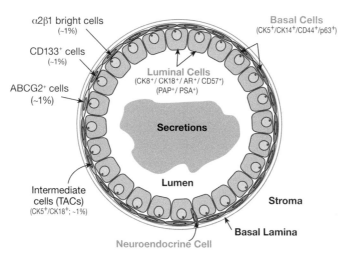

Figure 2–1: Prostatic glands are ductal structures composed of phenotypically and functionally distinct cells. Prostate epithelial cells include basal cells, located adjacent to the basal lamina, and cuboidal exocrine luminal cells, above and facing the lumen. CK5$^+$/CK14$^+$/CD44$^+$/p63$^+$ basal cells also include $\alpha2\beta1^{high}$, ABCG2$^+$, and/or CD133$^+$ subsets that may be partially overlapping. CK8$^+$/CK18$^+$/CD57$^+$/AR$^+$ luminal cells function to produce prostatic secretions, including prostate-specific antigen (PSA), probasin (Pb), and prostatic acid phosphatase (PAP). Intermediate cells (located in either the basal or luminal layers) express both basal and luminal markers (e.g., CK5$^+$CK18$^+$) and possibly represent transit-amplifying cells (TACs), and as such, they may co-express prostate stem cell antigen (PSCA), a marker of prostate progenitor cells. The cell surface marker Sca-1 (Stem Cell Antigen–1) is broadly expressed in mouse prostate, but in the adult, it marks primitive, multipotent stem/progenitor cells co-expressing integrin α6 (CD49f). Neuroendocrine cells, interspersed between the basal and luminal layers, may be identified as chromograninA$^+$/synaptophysin$^+$ cells. *See color plates.*

Although the developmental pathways and cell lineages of the prostate are relatively well understood, the precise identity and localization of the prostatic SCs have been scarcely elucidated, especially in humans. The expression of molecular markers permits the discrimination of basal (e.g., CK5$^+$, CK14$^+$, CD44$^+$, and p63$^+$) versus luminal (e.g., CK8$^+$, CK18$^+$, CD57$^+$, PSA$^+$, and AR^{high}) cells[39,45–48] (see Figure 2–1). Additional cell types present at lower frequency may also be distinguished and include neuroendocrine cells (e.g., chromogranin$^+$, synaptophysin$^+$, AR$^-$) and intermediate cell types that coexpress both basal and luminal cell markers (e.g., CK5$^+$, CK18$^+$) (see Figure 2–1). The exact interrelationships between these lineages is subject to speculation. Various proposed models include a linear relationship, with renewing, slow-cycling prostatic SCs residing in the basal layer, giving rise to differentiated basal and luminal cells (and, possibly, neuroendocrine cells) via rapidly proliferating transit-amplifying cells (see Figure 2–2). Alternatively, distinct basal and luminal SCs could exist separately, each giving rise to differentiated counterparts in their respective cell layers. Some observations supporting the existence of luminal SCs include evidence of long-lived, quiescent LRCs in both the basal and luminal cell layers of the mouse prostate[9] as well as evidence that p63$^{-/-}$ cells can generate prostatic tissue containing luminal-like cells even in the absence of phenotypically normal basal cells.[49] However, the prevailing view is that the predominance of self-renewing,

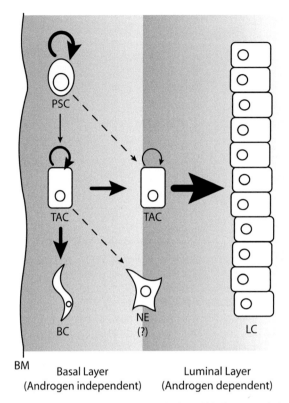

Basal Layer
(Androgen independent)

Luminal Layer
(Androgen dependent)

Figure 2–2: A hierarchical model of possible lineage relationships between cells in the normal adult prostate. Slow-cycling prostate stem cells (PSCs), residing in the basal layer adjacent to the basement membrane, possess extensive self-renewal abilities (curved arrow) and multipotency. PSCs divide to give rise to highly proliferative TACs with reduced renewal potential. TACs generate lineage-restricted terminally differentiated basal cells and luminal cells. Alternative origins are also possible; for example, distinct luminal TACs may arise via basal-layer TACs or directly from PSCs (dashed arrow). The origin of neuroendocrine cells is currently unknown, but these cells may derive from PSC or TAC cells or, alternatively, they may originate from neural crest cells during development.

stem-like prostatic cells resides in the basal layer, as we shall describe subsequently.

Purification and subsequent characterization of cells expressing distinct surface markers have provided further evidence for the existence of SCs in the prostatic epithelium. For example, in human prostate, very rare CD133[+]/α2β1[+] cells localized within the basal cell layer have been shown to possess a higher proliferative capacity in vitro and the ability to produce differentiated CK18[+] luminal cells on transplantation (with human stromal cells) into the flanks of athymic nude mice in vivo.[10] A murine-restricted cell surface marker Sca-1 (Stem Cell Antigen–1) has been shown to mark slow-cycling multipotent LRCs enriched in the mouse prostate tubules proximal to the urethra.[50,51]

The development of tissue recombination techniques by Cunha and Lung[52] has proven to be a powerful methodology to analyze the tissue reconstitution abilities of prostate cells.[40,53] Dissociated prostate epithelial cell populations are combined with urogenital sinus mesenchyme (UGSM) from mid-gestation rat (or mouse) embryos and transplanted under the kidney capsule in immunocompromised recipient mice. Stromal-epithelial cell signaling induces growth and differentiation of the epithelial cells, permitting analysis of candidate prostatic SCs and/or loss-of-function effects of particular genes. Using this approach, normal murine prostatic epithelial SCs have been definitively identified recently as Sca-1$^+$CD49f$^+$ (integrin α6) basal cells, accounting for virtually all of the clonogenic potential of prostatic epithelial cells in vitro and the ability to recapitulate prostatic tubules when transplanted as tissue recombinants in vivo.[11,37] Remarkably, these tubules are of a clonal origin, providing compelling evidence of the existence (and identity) of normal mouse pluripotent prostate stem cells.

PROSTATE CANCER (PCa)

Prostatic hyperproliferative diseases occur with increasing frequency in older males. These include benign prostate hyperplasia (BPH), a nonmalignant expansion of prostatic stromal and epithelial cells; prostatic intraepithelial neoplasia (PIN), an early malignant expansion of prostatic epithelial cells; and the most detrimental, invasive adenocarcinoma with metastatic potential. Although the etiology of PCa remains poorly understood, clearly a loss of homeostasis (i.e., aberrant cellular expansion) is fundamental to the manifestation of the disease. Inflammation, diet, and genetic predisposition have been implicated in the natural activation of prostate cell turnover due to cell death and subsequent proliferation during regeneration.[43] Importantly, PCa occurs in human males with high incidence and is generally multifocal, with numerous distinct lesions often present in the same gland. The susceptibility of the prostate and other epithelial tissues with regenerative properties, such as the skin, intestine, and breast, to tumor development suggests that renewing cells may be particularly amenable to transformation.

PCa, like many other epithelial tumors, has been proposed to be an SC disease such that relatively rare tumorigenic stem-like cells potentiate prostate tumorigenesis.[15,36,37] Furthermore, in consideration that normal prostatic SCs are androgen-independent (see previous discussion and Figure 2–2), prostate CSCs could survive conventional androgen ablation therapies used to treat PCa patients with invasive and metastatic tumors. Unfortunately, as is the case with breast cancer, hormone ablation therapies are generally not curative, and although tumors often dramatically regress short term, the tumors usually recur as hormone-refractory disease.[36,37,43] Therefore prostate CSCs are hypothesized to contribute to tumor initiation, tumor maintenance, and disease progression. Presuming PCa stem/progenitor cells exist, how can these cells be identified, characterized, and ultimately targeted?

PROSTATE CSCs

Recently, several laboratories have prospectively isolated and characterized candidate PCa stem/progenitor cells from both mouse and human prostates. The initial approach targeted the cytotoxic drug–effluxing SP and used the fluorescence-activated cell sorter (FACS) to separate these cells from dissociated prostate tissues. One study[54] utilized tissue samples from patients with BPH, and the results revealed that approximately 1.5% of the total population presented as SP (higher than is usually present in the hematopoietic lineage) and contained a mixture of proliferative cells and smaller quiescent cells. Integrin α2 was found to be enriched in the SP fraction; however, the size and granularity of the positive cells indicated that this marker was present on the cell surface of both actively cycling and quiescent cells, leading the authors to propose that this marker is not exclusive to prostatic SCs and may be expressed by transit-amplifying cells, as well.[54]

Subsequently, our group performed a further analysis of SP cells isolated from the androgen-dependent LAPC9 PCa xenograft tumors (derived from a human prostate carcinoma bone metastasis[55]), representing less than 1% of the bulk tumor mass, and found an enrichment in the tumorigenicity of these cells transplanted subcutaneously (s.c.) into NOD/SCID mice, as compared to the non-SP cells.[34] However, the potential toxicity of the Hoechst dye to the non-SP cells may confound such studies, and so the MDR transporter ABCG2 was also assayed (by immunofluorescence cell sorting) as a candidate marker for prostate CSCs. Interestingly, although ABCG2$^+$ Du145 cells exhibit higher cloning efficiency in vitro at early time points, ABCG2$^-$ cells actually gave rise to a higher frequency of large colonies later. Taken together with observations that ABCG2$^+$ and ABCG2$^-$ Du145 cells were similarly tumorigenic when transplanted s.c. into immuno-compromised recipient mice, these findings suggest that ABCG2 may not be the key mediator of the SP accounting for the increased tumorigenic potential of these cells. Gene expression profiling in this study was consistent with identification of ABCG2$^+$ cells as tumor progenitor cells, rather than CSCs.[34] However, researchers in a separate study found similar gene expression profiles between the SP and ABCG2$^+$ prostate cells, leading authors to propose that both populations may represent prostatic SCs.[56]

Normal human prostatic basal cells express the cell adhesion molecule CD44.[46] CD44 isoforms, or splice variants, have also been recently evidenced to be a marker of cancer stem/progenitor cells in a variety of tissues, including the breast and prostate.[24,29,57] Immunofluorescence-based cell sorting for CD44$^+$ cells from xenograft tumors, such as LAPC4 (lymph node metastasis derived), LAPC9, and Du145, enriched for highly tumorigenic and often metastatic PCa cells,[29] with CD44-expressing cells ranging widely in frequency from 1% up to 20% of the total cells. In addition to exhibiting higher tumor incidence and shorter latency when injected s.c. in vivo, CD44$^+$ Du145 cells plated at clonal density (in two dimensions on plastic) proliferated more extensively than did CD44$^-$ cells. Of particular interest, CD44$^+$ (AR$^-$) cells from LAPC4 and LAPC9 were shown to possess sphere-forming abilities and multipotency, as evidenced by the ability of these cells to give rise to CD44$^-$ and/or AR$^+$ cells. Taken together with observations that these cells express mRNA for some potential renewal genes, such as β-catenin, Smoothened (a signal transducer in the Sonic Hedgehog pathway) and Bmi (a member of the Polycomb group family of transcriptional repressors) were enriched in the CD44[+] population, and these findings suggest that CD44[+] PCa cells are stem-like cells. However, as CD44[−] cells, at high cell numbers and long latencies, can give rise to tumors containing both CD44[−] and CD44[+] cells, it is possible that a minor subset of CD44[−] cells are actually more primitive CSCs than the CD44[+] cells, although relative contributions of contaminating cell populations (based on the purity of the cell preparation) cannot be eliminated as a contributing factor.

Further elucidation of the human PCa lineage hierarchy has analyzed the collagen-interacting integrins ($\alpha2\beta1$) and the previously identified normal human prostate SC marker CD133,[10] a marker that has subsequently been shown to be expressed by PCa cells in situ.[58] Collins and colleagues[28] adapted their protocol to primary prostate tumor specimens and showed that the in vitro proliferative potential of CD44[+]/$\alpha2\beta1^{high}$/CD133[+] cells (less than 1% of the bulk population) is significantly higher than the negative or single positive cells (CD44[+]/$\alpha2\beta1^{high}$/CD133[+] > CD44[+]/$\alpha2\beta1^{high}$/CD133[−] $\gg$ CD44[+]/$\alpha2\beta1^{low}$/CD133[−]). Furthermore, these cells were shown to possess enhanced secondary cloning efficiency, suggesting that the CD133[+] fraction contains long-term renewing abilities, and gave rise to a mixed population of cells in the presence of androgen, therefore also exhibiting the SC feature of multipotency.[28] However, the tumorigenic potential of these cells was not reported, and in independent studies using the LAPC4/LAPC9 xenograft models, we have found similar tumorigenicities between CD44[+]/$\alpha2\beta1^{high}$ and CD44[+]/$\alpha2\beta1^{low}$ cells.[59] A hierarchical model has been proposed in which the cells with the highest proliferative and renewal potential include the SP and CD133[+] cells partially overlapping CD44[+] cells and marking putative prostate CSCs. In this schematic, expression of CD44 extends farther downstream in the lineage hierarchy to include $\alpha2\beta1^{high}$ and ABCG2[+] progenitor cells, followed by the fully differentiated CD44[−] (and AR[+]) PCa cells with the lowest clonogenic and tumorigenic potentials[29,36] (see Figure 2–3).

Long-term cultured human prostate tumor cells (and xenografts) have been evidenced to contain stem-like cells with enhanced cloning efficiency,[60] clonogenicity (both in two and three dimensions), and tumorigenicity.[36,61] We have found that prostate tumors initiated from holoclones can be serially passaged and express higher levels of stemness markers such as CD44 and β-catenin.[61] In separate experiments, hTERT immortalized HPCa-derived epithelial cell lines have recently been developed that exhibit SC characteristics, including tumor reconstitution (i.e., multipotency), expression of the early progenitor cell surface marker CD44, and expression of mRNA encoding CD133 as well as the ESC cell fate regulatory transcription factor Nanog.[62]

SELF-RENEWAL MACHINERY AND THE MOLECULAR MECHANISMS REGULATING CSC CELL FATE

Considering that tumor-initiating and tumor-maintaining cells possess certain attributes of SCs, what molecular mechanisms account for the acquisition (or

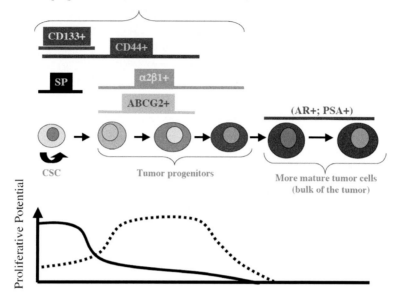

Figure 2–3: A tumorigenic hierarchy provides evidence of prostatic cancer stem cells (CSCs) in xenograft tumors. Prostate CSCs are relatively rare tumor-(re)initiating cells that can be prospectively purified and assayed in vitro for proliferative potential (e.g., clonal and/or clonogenic growth) and in vivo for tumorigenicity by transplantation into immuno-compromised recipient mice. In this model, CSCs, possessing the highest proliferative capacity and self-renewal (solid line), and tumor progenitors, possessing the highest proliferation rate (dashed line), are the most clonogenic and tumorigenic, whereas the bulk of the tumor, consisting of differentiated luminal-like cells, are non-tumorigenic. *See color plates.*

maintenance) of the SC state? Over expression of classical self-renewal genes has proven oncogenic in some systems, such as the ectopic expression of Oct4,[63,64] constitutively active β-catenin,[65] and Sonic Hedgehog.[66] Indeed, many of the cell-signaling pathways known to regulate normal SC self-renewal, including Wnt/β-catenin, Sonic Hedgehog, Notch, and PTEN, have also been shown to play pivotal roles in tumorigenesis, possibly by maintaining CSC self-renewal.[14,22,67]

A panoply of genetic mutations and epigenetic alterations, including those that affect renewal processes, have been evidenced to contribute to PCa, and these alterations can occur in innumerable combinations. Interestingly, transduction of Sca-1+ murine prostate SC-enriched cells with constitutively active AKT (a critical regulatory protein of the AKT/PTEN signaling axis), followed by recombination of these cells with UGSM and implantation under the kidney capsule, was sufficient to initiate tumorigenesis, as indicated by the presence of PIN lesions and prostate carcinoma in situ within the enlarged grafts.[51]

In separate studies, conditional knockout of PTEN has been shown to lead to metastatic PCa, and although most of the prostate tumor cells in this transgenic system undergo apoptosis in response to castration and regress, some androgen-independent cells survive and continue to proliferate,[68] reminiscent of human hormone-refractory disease. Furthermore, PTEN deletion leads to the expansion of stem/progenitor cells in these PTEN-null prostate tumors, including the Sca-1$^+$ population that overlaps with basal cells (p63$^+$/Bcl2$^+$) and transit-amplifying cells (CK5$^+$/CK8$^+$).[69] These findings suggest that PTEN negatively regulates the proliferation and, possibly, the self-renewal of prostate stem/progenitor cells.

Other potential regulators of proliferation and cell fate (i.e., self-renewal) evidenced to play a role in PCa may include Wnt/β-catenin,[70] Notch/Jagged,[71,72] Sonic Hedgehog,[73,74] and Stat3.[75,76] Additional novel self-renewal molecules regulating prostate tumorigenesis will, undoubtedly, be revealed by future studies.

THE CELL OF ORIGIN OF PCa

Although the majority of human prostate tumors are composed of differentiated luminal-like cells (e.g., CD44$^-$/AR$^+$), these cells appear to have reduced clonogenicity in vitro and tumorigenicity in vivo.[28,29] On the contrary, a subset of multipotent cells expressing stem/progenitor cell markers (e.g., CD44 and/or CD133) has been shown to enrich for clonogenic and tumorigenic cells.[28,29] These observations suggest origination within a primitive cell. On the other hand, because these phenotypic characteristics (i.e., expression of cell surface markers) can be acquired, these experimental findings do not sufficiently address the cell of origin of prostate tumorigenesis. Genetically tractable mouse models of PCa will likely provide additional evidence of tumorigenic transformation in prostatic stem/progenitor cells. For example, PTEN deletion in both basal and luminal cells leads to the selective expansion of basal-like Sca-1$^+$ stem/progenitor cells coincident with tumor initiation,[69] suggesting that primitive cells within the basal layer may be the cell of origin in this model system. Further experiments, such as lineage tracking, will be required to further elucidate the nature of the cell of origin (or, more likely, *cells* of origin) of PCa.

CONCLUSION AND SIGNIFICANCE

PCa is a disease that progresses from an androgen-dependent to an androgen-independent state. Normal prostatic SCs are known be insensitive to androgen withdrawal, and it appears that prostatic CSCs are also able to survive androgen ablation. Consequently, these cells represent critical clinical therapeutic targets in the successful treatment of hormone-refractory disease, and novel methods to selectively target prostate CSCs are urgently needed. The most effective target may, in fact, be renewing stem-like cells *regardless* of the cell of origin, and successful therapy could potentially cure PCa by reestablishing tissue homeostasis such that a balance among proliferation, cell death, and differentiation is reacquired. It is our hope that the current investigations into PCa stem/progenitor cells will shed light on the etiology of the disease and reveal unique molecular mechanisms by which these tumor-initiating, tumor-maintaining, and often metastasis-mediating malignant cells may be targeted in the clinic.

ACKNOWLEDGMENTS

We thank the MDACC Science Park–Research Facility and Animal Cores for technical assistance; Drs. R. Fagin, S. Pickett, J. Waxman, and L. Dalton for providing primary HPCa samples; and other members of the Tang lab for support and helpful discussions. This work was supported in part by grants from NIH (R01-AG023374, R01-ES015888, and R21-ES015893-01A1), the American Cancer Society (RSG MGO-105961), the U.S. Department of Defense (W81XWH-07-1-0616), the Prostate Cancer Foundation, and the Elsa Pardee Foundation (D.G.T) and by two Center Grants (CCSG-5 P30 CA166672 and ES07784). C.R.J. was supported in part by a postdoctoral fellowship from NIH and the American Urological Association.

REFERENCES

1. Welm, B.E., Tepera, S.B., Venezia, T., Graubert, T.A., Rosen, J.M., and Goodell, M.A. (2002) Sca-1^{pos} cells in the mouse mammary gland represent an enriched progenitor cell population. *Dev. Biol.* **245**, 42–56.
2. Dontu, G., Abdullah, W.M., Foley, J.M., Jackson, K.W., Clarke, M.F., Kawamura, M.J., and Wicha, M. (2003) In vitro propagation and transcriptional profiling of human mammary stem/progenitor cells. *Genes Dev.* **17**, 1253–1270.
3. Clarke, R.B., Spence, K., Anderson, E., Howell, A., Okano, H., and Potten, C.S. (2005) A putative human breast stem cell population is enriched for steroid receptor-positive cells. *Dev. Biol.* **277**, 443–456.
4. Shackleton, M., Vaillant, F., Simpson, K.J., Stingl, J., Smyth, G.K., Asselin-Labat, M., Wu, L., Lindeman, G.J., and Visvader, J.E. (2006) Generation of a functional mammary gland from a single stem cell. *Nature* **439**, 84–88.
5. Cotsarelis, G., Sun, T.T., and Lavker, R.M. (1990) Label-retaining cells reside in the bulge of the pilosebaceous unit: implications for follicular stem cells, hair cycle, and skin carcinogenesis. *Cell* **61**, 1329–1337.
6. Morris, R.J., Liu, Y., Marles, L., Yang, Z., Trempus, C., Li, S., Lin, J.S., Sawicki, J.A., and Cotsarelis, G. (2004) Capturing and profiling adult hair follicle stem cells. *Nat. Biotech.* **22**, 411–417.
7. Ohyama, M., Terunuma, A., Tock, C.L., Radonovich, M.F., Pise-Masison, C.A., Hopping, S.B., Brady, J.N., Udey, M.C., and Vogel, J.C. (2006) Characterization and isolation of stem-cell enriched human hair follicle bulge cells. *J. Clin. Invest.* **116**, 249–260.
8. Kim, S.J., Cheung, S., and Hellerstein, M.K. (2004) Isolation of nuclei from label-retaining cells and measurement of their turnover rates in rat colon. *Am. J. Physiol. Cell. Physiol.* **286**, C1464–C1473.
9. Tsujimura, A., Koikawa, Y., Salm, S., Takao, T., Coetzee, S., Moscatelli, D., Shapiro, E., Lepor, H., Sun, T.T., and Wilson, E.L. (2002) Proximal location of mouse prostate epithelial stem cells: a model of prostatic homeostasis. *J. Cell. Biol.* **157**, 1257–1265.
10. Richardson, G.D., Robson, C.N., Lang, S.H., Maitland, N.J., and Collins, A.T. (2004) CD133, a novel marker for human prostatic epithelial stem cells. *J. Cell. Sci.* **117**, 3539–3545.
11. Lawson, D.A., Xin, L., Lukas, R.U., Cheng, D., and Witte, O.N. (2007) Isolation and functional characterization of murine prostate stem cells. *Proc. Natl. Acad. Sci. U. S. A.* **104**, 181–186.

12. Chambers, I., and Smith, A. (2004) Self-renewal of teratocarcinoma and embryonic stem cells. *Oncogene* **23**, 7150–7160.

13. Tumbar, T., Guasch, G., Greco, V., Blanpain, C., Lowry, W.E., Rendl, M., and Fuchs, E. (2004) Defining the epithelial stem cell niche in skin. *Science* **303**, 359–363.

14. Miller, S.J., Lavker, R.M., and Sun, T.T. (2005) Interpreting epithelial cancer biology in the context of stem cells: tumor properties and therapeutic implications. *Biochim. Biophys. Acta* **1756**, 25–52.

15. Lam, J.S., and Reiter, R.E. (2006) Stem cells in prostate and prostate cancer development. *Urol. Oncol.* **24**, 131–140.

16. Fuchs, E., Tumbar, T., and Guasch, G. (2004) Socializing with the neighbors: stem cells and their niche. *Cell* **116**, 769–778.

17. Moore, K.A., and Lemischka, I.R. (2006) Stem cells and their niches. *Science* **311**, 1880–1885.

18. Reya, T., Morrison, S.J., Clarke, M.F., and Weissman, I.L. (2001) Stem cells, cancer, and cancer stem cells. *Nature* **414**, 105–111.

19. Al-Hajj, M., and Clarke, M.F. (2004) Self-renewal and solid tumor stem cells. *Oncogene* **23**, 7274–7282.

20. Bruce, W.R., and Van Der Gaag, H. A quantitative assay for the number of murine lymphoma cells capable of proliferation *in vivo*. *Nature* **199**, 79–80.

21. Park, C.H., Bersagagel, D.E., and McCulloch, E.A. (1971) Mouse myeloma tumor stem cells: a primary cell culture assay. *J. Natl. Cancer Inst.* **46**, 411–422.

22. Wicha, M.S., Liu, S., and Dontu, G. (2006) Cancer stem cells: an old idea – a paradigm shift. *Cancer Res.* **66**, 1883–1890.

23. Bonnet, D., and Dick, J.E. (1997) Human acute myeloid leukemia is organized as a hierarchy that originates from a primitive hematopoietic cell. *Nature Med.* **3**, 730–737.

24. Al-Hajj, M., Wicha, M.S., Benito-Hernandez, A., Morrison, S.J., and Clarke, M.F. (2003) Prospective identification of tumorigenic breast cancer cells. *Proc. Natl. Acad. Sci. U. S. A.* **100**, 3983–3988.

25. Singh, S.K., Hawkins, C., Clarke, I.D., Squire, J.A., Bayani, J., Hide, T., Henkelman, R.M., Cusimano, M.D., and Dirks, P.B. (2004) Identification of human brain tumor initiating cells. *Nature* **432**, 396–401.

26. O'Brien, C.A., Pollett, A., Gallinger, S., and Dick, J.E. (2007) A human colon cancer cell capable of initiating tumour growth in immunodeficient mice. *Nature* **445**, 106–110.

27. Ricci-Vitiani, L., Lombardi, D.G., Pilozzi, E., Biffoni, M., Todaro, M., Peschle, C., and De Maria, R. (2007) Identification and expansion of human colon-cancer-initiating cells. *Nature* **445**, 111–115.

28. Collins, A.T., Berry, P.A., Hyde, C., Stower, M.J., and Maitland, N.J. (2005) Prospective identification of tumorigenic prostate cancer stem cells. *Cancer Res.* **65**, 10946–10951.

29. Patrawala, L., Calhoun, T., Schneider-Broussard, R., Li, H., Bhatia, B., Tang, S., Reilly, J.G., Chandra, D., Zhou, J., Claypool, K., Coghlan, L., and Tang, D.G. (2006) Highly purified CD44+ prostate cancer cells from xenograft human tumors are enriched in tumorigenic and metastatic progenitor cells. *Oncogene* **25**, 1696–1708.

30. Singh, S.K., Clarke, I.D., Terasaki, M., Bonn, V.E., Hawkins, C., Squire, J., and Dirks, P.B. (2003) Identification of a cancer stem cell in human brain tumors. *Cancer Res.* **63**, 5821–5828.

31. Hemmati, H.D., Nakano, I., Lazareff, J.A., Masterman-Smith, M., Geschwind, D.H., Bronner-Fraser, M., and Kornblum, H.I. (2003) Cancerous stem cells can arise from pediatric brain tumors. *Proc. Natl. Acad. Sci. U. S. A.* **100**, 15178–15183.

32. Kondo, T., Setoguchi, T., and Taga, T. (2004) Persistence of a small subpopulation of cancer stem-like cells in the C6 glioma cell line. *Proc. Natl. Acad. Sci. U. S. A.* **101**, 781–786.

33. Hirschmann-Jax, C., Foster, A.E., Wulf, G.G., Nuchtern, J.G., Jax, T.W., Gobel, U., Goodell, M.A., and Brenner, M.K. (2004) A distinct "side population" of cells with high drug efflux capacity in human tumor cells. *Proc. Natl. Acad. Sci. U. S. A.* **101**, 14228–14233.

34. Patrawala, L., Calhoun, T., Schneider-Broussard, R., Zhou, J.-J., Claypool, K., and Tang, D.G. (2005) Side population is enriched in tumorigenic, stem-like cancer cells, whereas ABCG2$^+$ and ABCG2$^-$ cancer cells are similarly tumorigenic. *Cancer Res.* **65**, 6207–6219.

35. Naumov, G.N., Townson, J.L., MacDonald, I.C., Wilson, S.M., Bramwell, V.H.C., Groom, A.C., and Chambers, A.F. (2003) Ineffectiveness of doxorubicin treatment on solitary dormant mammary carcinoma cells or late-developing metastases. *Breast Cancer Res.* **82**, 199–206.

36. Tang, D.G., Patrawala, L., Calhoun, T., Bhatia, B., Schneider-Broussard, R., Choy, G., and Jeter, C. (2007) Prostate cancer stem/progenitor cells: identification, characterization and implications. *Mol. Carcinogen.* **46**, 1–14.

37. Lawson, D.A., and Witte, O.N. (2007) Stem cells in prostate cancer initiation and progression. *J. Clin. Invest.* **117**, 2044–2050.

38. Abate-Shen, C., and Shen, M. (2000) Molecular genetics of prostate cancer. *Genes Dev.* **14**, 2410–2434.

39. Wang, Y., Hayward, S.W., Cao, M., Thayer, K., and Cunha, G.R. (2001) Cell differentiation lineage in the prostate. *Differentiation* **68**, 270–279.

40. Xin, L., Ide, H., Kim, Y., Dubey, P., and Witte, O. (2003) *In vivo* regeneration of murine prostate from dissociated cell populations of postnatal epithelia and urogenital sinus mesenchyme. *Proc. Natl. Acad. Sci. U. S. A.* **100**, 11896–11903.

41. Aumuller, G., Leonhardt, M., Janssen, M., Konrad, L., Bjartell, A., and Abrahamsson, P.A. (1999) Neurogenic origin of human prostate endocrine cells. *Urology* **53**, 1041–1048.

42. Rizzo, S., Attard, G., and Hudson, D.L. (2005) Prostate epithelial stem cells. *Cell Prolif.* **38**, 363–374.

43. Litvinov, I., DeMarzo, A.M., and Isaacs, J. (2003) Is the Achilles' heel for prostate cancer therapy a gain of function in androgen receptor signaling? *J. Clin. Endocrin. Metab.* **88**, 2972–2982.

44. English, H.F., Santen, R.J., and Isaacs, J.T. (1987) Response of glandular versus basal rat ventral prostatic epithelial cells to androgen withdrawal and replacement. *Prostate* **11**, 229–242.

45. Sherwood, E.R., Theyer, G., Steiner, G., Berg, L.A., Kozlowski, J.M., and Lee, C. (1991) Differential expression of specific cytokeratin polypeptides in the basal and luminal epithelia of the human prostate. *Prostate* **18**, 303–314.

46. Liu, A.Y., True, L.D., LaTray, L., Nelson, P.S., Ellis, W.J., Vessella, R.L., Lange, P.H., Hood, L., and Van Den Engh, G. (1997) Cell-cell interaction in prostate gene regulation and cytodifferentiation. *Proc. Natl. Acad. Sci. U. S. A.* **94**, 10705–10710.

47. Signoretti, S., Waltregny, D., Dilks, J., Isaac, B., Lin, D., Garraway, L., Yang, A., Montironi, R., McKeon, F., and Loda, M. (2000) p63 is a prostate basal cell marker and is required for prostate development. *Am. J. Pathol.* **157**, 1769–1775.

48. Collins, A.T., Habib, F.K., Maitland, N.J., and Neal, D.E. (2001) Identification and isolation of human prostate epithelial stem cells based on alpha(2)beta(1)-integrin expression. *J. Cell Sci.* **114**, 3865–3871.

49. Kurita, T., Medina, R.T., Mills, A.A., and Cunha, G.R. (2004) Role of p63 and basal cells in the prostate. *Development* **131**, 4955–4964.

50. Burger, P.E., Xiong, X., Coetzee, S., Salm, S.N., Mascatelli, D., Goto, K., and Wilson, E.L. (2005) Sca-1 expression identifies stem cells in the proximal region of prostatic ducts with high ability to reconstitute prostatic tissue. *Proc. Natl. Acad. Sci. U. S. A.* **102**, 7180–7185.

51. Xin, L., Lawson, D.A., and Witte, O.N. (2005) The Sca-1 cell surface marker enriches for a prostate-regenerating cell population that can initiate prostate tumorigenesis. *Proc. Natl. Acad. Sci. U. S. A.* **102**, 6942–6947.

52. Cunha, G.R., and Lung, B. (1978) The possible influence of temporal factors in androgenic responsiveness of urogenital tissue recombinants from wild-type and androgen-insensitive (Tfm) mice. *J. Exp. Zool.* **205**, 181–193.

53. Hayward, S.W., Haughney, P.C., Rosen, M.A., Greulich, K.M., Weier, H.G., Dahiya, R., and Cunha, G.R. (1998) Interactions between human prostatic epithelium and rat urogenital sinus mesenchyme in a tissue recombination model. *Differentiation* **63**, 131–140.

54. Bhatt, R.I., Brown, M.D., Hart, C.A., Gilmore, P., Ramani, V.A., George, N.J., and Clarke, N.W. (2003) Novel method for the isolation and characterisation of the putative prostatic stem cell. *Cytometry Part A* **54**, 89–99.

55. Craft, N., Chhor, C., Tran, C., Belldegrun, A., DeKernion, J., Witte, O.N., Said, J., Reiter, R.E., and Sawyers, C.L. (1999) Evidence for clonal outgrowth of androgen-independent prostate cancer cells from androgen-dependent tumors through a two-step process. *Cancer Res.* **59**, 5030–5036.

56. Pascal, L.E., Oudes, A.J., Petersen, T.W., Goo, Y.A., Walashek, L.S., True, L.D., and Liu, A.Y. (2007) Molecular and cellular characterization of ABCG2 in the prostate. *BMC Urol.* **7**, 6–18.

57. Patrawala, L., and Tang, D.G. (2008) CD44 as a functional cancer stem cell marker and therapeutic target. In: *Progress in Gene Therapy: Autologous and Cancer Stem Cell Gene Therapy*, vol. **3**, pp. 317–334 (Bertolotti, R., and Ozawka, K., eds). Hacxkensack, NJ: World Scientific.

58. Miki, J., Furusato, B., Li, H., Gu, Y., Takahashi, H., Egawa, S., Sesterhenn, I., McLeod, D., Srivastava, S., and Rhim, J.S. (2007) Identification of putative stem cell markers, CD133 and CXCR4, in hTERT-immortalized primary nonmalignant and malignant tumor-derived human prostate epithelial cell lines and in prostate cancer specimens. *Cancer Res.* **67**, 3153–3161.

59. Patrawala, L., Calhoun-Davis, T., Schneider-Broussard, R., and Tang, D.G. (2007) Hierarchical organization of prostate cancer cells in xenograft tumors: the CD44$^+$α2β1$^+$ cell population is enriched in tumor-initiating cells. *Cancer Res.* **67**, 6796–6805.

60. Locke, M., Heywood, M., Fawell, S., and Mackenzie, I.C. (2005) Retention of intrinsic stem cell hierarchies in carcinoma-derived cell lines. *Cancer Res.* **65**, 8944–8950.

61. Li, H., Chen, X., Calhoun-Davis, T., Claypool, K., and Tang, D.G. (2008) PC3 human prostate carcinoma cell holoclones contain self-renewing tumor-initiating cells. *Cancer Res.* **68**, 1820–1825.

62. Gu, G., Yuan, J., Wills, M., and Kasper, S. (2007) Prostate cancer cells with stem cell characteristics reconstitute the original human tumor in vivo. *Cancer Res.* **67**, 4807–4815.

63. Gidekel, S., Pizov, G., Bergman, Y., and Pikarsky, E. (2003) Oct-3/4 is a dose-dependent oncogenic fate determinant. *Cancer Cell* **4**, 361–370.

64. Hochedlinger, K., Yamada, Y., Beard, C., and Jaenisch, R. (2005) Ectopic expression of Oct-4 blocks progenitor-cell differentiation and causes dysplasia in epithelial tissues. *Cell* **121**, 465–477.

65. Gat, U., DasGupta, R., Degenstein, L., and Fuchs, E. (1998) De novo follicle morphogenesis and hair tumors in mice expressing truncated b-catenin in skin. *Cell* **95**, 605–614.

66. Oro, A.E., Higgins, K.M., Hu, Z., Boniface, J.M., Epstein, E.H., Jr., and Scott, M.P. (1997) Basal cell carcinomas in mice overexpressing Sonic Hedgehog. *Science* **276**, 817–821.

67. Rossi, D.J., and Weissman, I.L. (2006) Pten, tumorigenesis, and stem cell self-renewal. *Cell* **125**, 229–231.

68. Wang, S., Gao, J., Lei, Q., Rozengurt, N., Pritchard, C., Jiao, J., Thomas, G.V., Li, G., Roy-Burman, P., Nelso, P.S., Liu, X., and Wu, H. (2003) Prostate-specific deletion of the murine Pten tumor suppressor gene leads to metastatic prostate cancer. *Cancer Cell* **4**, 209–221.

69. Wang, S., Garcia, A.J., Wu, M., Lawson, D.A., Witte, O.N., and Wu, H. (2006) Pten deletion leads to the expansion of a prostatic stem/progenitor cell subpopulation and tumor initiation. *Proc. Natl. Acad. Sci. U. S. A.* **103**, 1480–1485.

70. Yardy, G.W., and Brewster, S.F. (2005) Wnt signaling and prostate cancer. *Prostate Cancer Prostatic Dis.* **8**, 119–126.

71. Santagata, S., Demichelis, F., Riva, A., Varambally, S., Hofer, M., Kutok, J.L., Kim, R., Tang, J., Montie, J.E., Chinnaiyan, A.M., Rubin, M.A., and Aster, J.C. (2004) JAGGED1 expression is associated with prostate cancer metastasis and recurrence. *Cancer Res.* **64**, 6854–6857.
72. Zayzafoon, M., Abdulkadir, S.A., and McDonald, J.M. (2004) Notch signaling and ERK activation are important for the osteomimetic properties of prostate cancer bone metastatic cell lines. *J. Biol. Chem.* **279**, 3662–3670.
73. Sanchez, P., Hernandez, A.M., Stecca, B., Kahler, A.J., DeGueme, A.M., Barrett, A., Beyna, M., Datta, M.W., Datta, S., and Altaba, A.R. (2004) Inhibition of prostate cancer proliferation by interference with SONIC HEDGEHOG-GLI1 signaling. *Proc. Natl. Acad. Sci. U. S. A.* **101**, 12561–12566.
74. Karhadkar, S.S., Bova, G.S., Abdallah, N., Dhara, S., Gardern, D., Maitra, A., Isaacs, J.T., Berman, D.M., and Beachy, P.A. (2004) Hedgehog signaling in prostate regeneration, neoplasia and metastasis. *Nature* **431**, 707–712.
75. Mora, L.B., Buettner, R., Seigne, J., Diaz, J., Ahmad, N., Garcia, R., Bowman, T., Falcone, R., Fairclough, R., Cantor, A., Muro-Cacho, C., Livingston, S., Karras, J., Pow-Sang, J., and Jove, R. (2002) Constitutive activation of Stat3 in human prostate tumors and cell lines: direct inhibition of Stat3 signaling induces apoptosis of prostate cancer cells. *Cancer Res.* **62**, 6659–6666.
76. Gao, L., Zhang, L., Hu, J., Li, F., Shao, Y., Zhao, D., Kalvakolanu, D., Kopecko, D., Zhao, X., and Xu, D. (2005) Down-regulation of Signal Transducer and Activator of Transcription 3 expression using vector-based small interfering RNAs suppresses growth of human prostate tumor *in vivo*. *Clin. Cancer Res.* **11**, 6333–6341.

3 Melanoma cancer stem cells

Alexander Roesch and Meenhard Herlyn
The Wistar Institute

BACKGROUND AND DEFINITIONS

Cutaneous melanoma is among the most aggressive types of human cancer, and if untreated, virtually every melanoma has the potential to metastasize.[1] While patients with locoregional disease and low tumor thickness can be cured in 90% of cases by surgery, the majority of patients with advanced disease die because of the inefficiency of current therapy regimens.[2]

Cutaneous melanoma is historically defined as a malignant tumor derived from the transformation and proliferation of epidermal melanocytes, enabling a stepwise progression from common melanocytic nevus to radial growth phase melanoma, vertical growth phase melanoma, and finally, metastatic disease.[3–5] However, recent data suggest that a considerable proportion – around 60% to 75% – of melanomas develop de novo, without any precursor lesions.[6,7] On the basis of these observations and repeated findings on melanoma heterogeneity,[8–10] an alternative hypothesis has been put forth in light of the emerging cancer stem cell (CSC) concept.[11] Mounting evidence suggests that

31

melanoma may arise from a multipotent CSC that is able to self-renew via asymmetric division, differentiate into diverse progenies, and drive continuous growth.[12–14] In this context, the term *melanoma stem cell* represents an operational definition indicating a multipotent tumor-initiating cell subset that – although monoclonal in origin – can give rise to a three-dimensional, heterogeneous progeny that caricatures the tissue of origin. According to this, melanoma would become functionally heterogeneous as a result of gradual differentiation of cells and not due to the coexistence of multiple genetic subclones resulting from independent somatic mutations.[15,16] Although parallels to normal melanocyte stem cells exist, the melanoma stem cell is provisionally defined by its stem cell–like properties, that is, the capability to self-renew and to transdifferentiate, and not by its cell of origin, which is still unknown to date (adult melanocyte stem cell, more differentiated progenitor cell, mature melanocyte, or a product of cell fusion).

PHYSIOLOGIC MAINTENANCE OF MELANOCYTES IN SKIN AND HAIR FOLLICLE: THE ADULT MELANOCYTE STEM CELL

Melanocytes, or rather, their progenitors, termed *melanoblasts*, represent migratory cells wandering from the neural crest to populate the basal layer of the early epidermis and the hair follicle.[17,18] Apparently, malignant melanoma cells recapitulate the migratory capacity of their immature progenitors, which may contribute to their highly metastatic potential.[19]

Recent data suggest how functioning and maintenance of normal melanocytes are further guaranteed after birth.[10,20] For example, studies using Dct-lacZ transgenic mice have allowed identification of putative melanocyte stem cells as a source of melanocyte maintenance within the lower permanent portion of the murine hair follicle.[21] Phenotypically immature, slow-cycling (dormant), and self-renewing melanocytic stem cells were found to be localized in the bulge area, also a well-characterized niche for epithelial adult stem cells. During the hair cycle, these melanocytic stem cells periodically differentiate into melanocyte precursor cells that migrate to the hair bulb, where they finally differentiate into mature melanocytes. Several studies have subsequently identified some candidate markers for these follicular melanocyte stem cells. Using single-cell cDNA amplification from dissected murine hair follicles, Osawa and colleagues obtained gene expression profiles indicating that most markers involved in melanocyte development, such as SOX10, Tyr, Tyrp1, Lef1, Kit, and Mitf, were absent in the melanocyte stem cells, whereas Pax3 and Dct were significantly expressed.[22] In contrast, melanocytes from the hair matrix at the bulb expressed all markers and therefore could be clearly distinguished.

According to the known functions of murine melanocyte stem cells, also in human hair follicles, melanocyte stem cells of the bulge area may represent an indefinite source of pigment cells. A corresponding population of cells with stem cell properties could recently be isolated from isolated human hair follicles.[23] Stemness of this newly detected adult stem cell population was confirmed by the cells' ability to self-renew in serial dilution assays and to differentiate into several tissues, including melanocytes, smooth muscle cells, and neurons, using tissue-specific culture conditions. Like their murine counterparts, these cells did not express Tyrp1 and Mitf. However, two embryonic stem cell markers, Oct4 and Nanog, were significantly expressed and were considered potential markers for identification of melanocyte stem cells in humans. Further studies are needed to elucidate the role of these cells during the human hair cycle and to explore their full stem cell potential, with further regard to a better understanding of abnormal stem cell functioning, for example, in melanoma.

In contrast to the situation in the epidermis, where pigmentation is maintained throughout life, in hair follicles, cyclic functioning of melanocyte stem cells can be temporally restricted, which becomes apparent when hair starts to gray. Using two different murine models of hair graying, the Bcl2 null mouse and the Mitf^{vit} mouse, Nishimura and colleagues[24] showed that hair graying is a result of incomplete melanocyte stem cell maintenance in the bulge region. From their studies with Bcl2 null mice, they concluded that Bcl2 provides a survival signal necessary between hair cycles to protect melanocyte stem cells from apoptosis during their transition from the active into the dormant state. Studies of the Mitf^{vit} mouse demonstrated that, without functional Mitf, melanocyte stem cells undergo ectopic, premature differentiation while residing in the bulge or lose their ability to migrate to the bulb area of the hair matrix before differentiation. For humans, comparable studies are still lacking.

It also remains unanswered whether there is another niche in the human skin apart from the bulge area that, for example, could explain the differences in pigmentation maintenance of skin and hair. Recent data on multipotent precursor cells isolated from murine neonatal skin or from human foreskin biopsies suggest that there might be a non–follicle-associated stem cell reservoir in the dermis, too.[25,26] However, it is also conceivable that no second epidermal niche exists. As demonstrated by Nishimura and colleagues,[21] transiently amplifying cells, which represent the direct progeny of melanocyte stem cells, can escape their follicular niche and migrate to the epidermis, where they can further differentiate into pigmented melanocytes. Interestingly, this finding is mirrored by clinical observations from patients with *vitiligo*, a depigmentation disorder resulting in the loss of epidermal melanocytes. After initial loss of skin pigmentation in vitiligo patients, a repigmentation, beginning from the hair follicles, can be observed during recovery, suggesting the existence of a migratory, follicle-derived cell pool that restocks the epidermis with melanin-producing melanocytes.[27]

EVIDENCE FOR A MELANOMA STEM CELL WITH SELF-RENEWAL AND TRANSDIFFERENTIATION PROPERTIES

During the past decades, indirect evidence has supported the presence of melanoma stem cells. First, melanomas show phenotypic heterogeneity both in vitro and in vivo, suggesting an origin from a cell with multilineage differentiation abilities. Melanomas retain their morphologic and biological plasticity,

Figure 3–1: Formation of melanoma spheres. **(a)** Using regular growth conditions for melanoma cells, dissociated cells from a primary melanoma sample adhere and develop a typical dendritic morphology. **(b)** If the dissociated cells are grown under human embryonic stem cell conditions, nonattaching cell clusters can be observed, which, finally, **(c)** grow to spheres.

despite repeated cloning.[8] Second, melanoma cells often express developmental genes such as Sox10, Pax3, and Mitf.[28,29] Melanomas also express the intermediate filament Nestin, which is associated with multiple stem cell populations.[30] Third, melanoma cells can differentiate into a wide range of cell lineages, including neural, mesenchymal, and endothelial cells. They frequently exhibit characteristics of neural lineages.[31]

Next to these indirect findings, recent scientific work has discovered direct evidence for the existence of melanoma stem cells. Applying growth conditions suitable to human embryonic stem cells, such as the use of a human embryonic stem cell medium, Fang and colleagues[12] found a subpopulation of melanoma cells propagating as nonadherent spheres in approximately 20% of metastatic melanomas, whereas in standard media, adherent monolayer cultures developed (Figure 3–1). The human embryonic stem cell medium was used because it selects stem cells, while differentiated cells rapidly die in response to the abundance of included mitogens.[12] Sphere formation of in vitro cultured cells had been supposed by different groups to be a common growth characteristic of stem cells, including neural crest–derived stem cells.[32–36]

Melanoma spheres can differentiate under appropriate culture conditions into multiple lineages, such as melanocytes, adipocytes, osteocytes, and chondrocytes, recapitulating the plasticity of neural crest stem cells. Multipotent melanoma spheroid cells persisted over several months after serial cloning in vitro and transplantation in vivo, indicating a stable capacity to self-renew. Interestingly, sphere cells were more tumorigenic than their adherent counterparts when grafted into mice. Finally, the authors found that the stemness criteria were significantly enriched in a small CD20-positive subpopulation of the spheres, indicating that CD20 might be a suitable surface marker for the identification of melanoma stem cells. This is interesting for two reasons: first, CD20 has been recently described by gene expression profiling as one of the top 22 markers defining aggressive melanomas,[37] and second, monoclonal antibodies against CD20 have already become a standard therapy in non-Hodgkin's lymphoma.[38]

Recent publications revealed another marker, CD133, to be commonly expressed in tumor-initiating populations from neuroectodermal tumors such as medulloblastoma and glioblastoma.[35,39] CD133, also known as prominin-1, is a cell surface glycoprotein first isolated from hematopoietic stem cells. Its function has not been fully established so far, but it may play a role in the regulation of membrane topology.[40] A tissue microarray study of a large number of human samples has now revealed that melanocytic tumors such as benign nevi, in situ melanomas, and advanced melanomas express CD133.[41] The authors found a stepwise increase in the proportion of CD133-positive cells from nevi to melanoma and melanoma metastases. Nevi revealed a more focal expression pattern, whereas melanoma and metastases showed a diffuse expression pattern. Using fluorescence-activated cell sorting (FACS) from freshly isolated melanoma cells, Monzani and colleagues[42] additionally demonstrated that the CD133-positive subpopulation represents less than 1% of the total tumor mass of melanoma, a finding consistent with designated stem cell subpopulations from other tissues. Like the CD20-positive population defined by Fang and colleagues,[12] CD133-positive melanoma cells revealed an increased tumorigenicity when injected into nonobese diabetic/severe combined immunodeficiency disease (NOD/SCID) mice. In contrast to the CD20-positive subpopulation, in Monzani and colleagues' report, CD133-positive cells were more present in the adherent fraction of the analyzed melanoma cell line compared to the floating subpopulation. However, because stringent experiments on self-renewal and the transdifferentiation capacity of CD133-positive melanoma cells are lacking, the role of CD133 as a potential melanoma stem cell marker remains elusive.

Another common criterion for identification of stem cells is the ability to efflux Hoechst 33342 dye and thus to exhibit low fluorescence in FACS analysis.[43] Using flow sorting of three melanoma cell lines isolated from lymph node metastases, Grichnik and colleagues[14] detected a tiny subpopulation (also called the side population) of small-sized cells with both high Hoechst 33342 efflux and the ability to give rise to cells of different morphology when cultured in vitro. Cells from this subpopulation additionally showed a low proliferation rate but a high ability to self-renew. Noteworthy was the finding that the premelanosomal marker gp100 and the known stem cell marker Nestin were more highly expressed in Hoechst 33342–low cells, whereas nerve growth factor receptor (NGFR) better labeled large, Hoechst 33342–high cells. The authors admitted that marker expression levels were highly variable and that the results could not be replicated under all culture conditions. However, the morphologic and antigenetic heterogeneity of the progeny remained stable after repeated single-cell cloning of Hoechst 33342–low cells. This strengthens the assumption that melanoma is based on a mutant stem cell giving rise to a phenotypically but not genetically heterogeneous progeny and is not based on the accumulation of different mutant clones (which would have been indicated by expansion of a uniform cell population after single-cell cloning). This hypothesis has been lately reinforced by clinical observations. Wang and colleagues[44] reported an interesting case of a patient with metastatic cutaneous melanoma who had experienced several recurrences over a decade of follow-up. From each recurrence, biopsies were taken and cell lines established. Using karyotyping and comparative genomic hybridization (CGH), the authors identified consistent genetic traits in spite of divergent phenotypes, suggesting that all metastases were derived from the same clone of origin and that, after metastatic nidation, a phenotypically variable progeny arose. Some years before this study, clonal progression of primary cutaneous melanoma had been shown for the transition from the radial growth phase into vertical-growth-phase melanoma (CGH)[45,46] and the development of in-transit metastases (loss of heterozygosity).[47]

However, the question of which definite cell type initially acquires the essential tumor-initiating genetic hit has not been answered by any of the studies reported so far. If cancer arises from a series of genetic mutations, stem cells would have more opportunity to accumulate mutations because of their long life span. Adult melanocyte stem cells may be the ideal target for accumulating mutations, but the rarity of adult melanocyte stem cells may counter this theory because of the low probability that they could be targeted by mutations, for example, due to their exposition to potential carcinogens like ultraviolet irradiation. The relative abundance of transient amplifying progenitor cells derived from stem cells and their ability to retain a partial self-renewal capacity makes them more likely candidates for initial transforming events. Finally, it is also conceivable that mutated somatic cells can fuse with normal stem cells, thereby generating cancer stem cells.[48]

MOLECULAR SIGNALING PATHWAYS THAT MAINTAIN STEMNESS

Self-renewal and quiescence of normal stem cells are regulated in various organ systems by the same key signaling pathways that are also found deregulated in the progression and metastases of many tumors, including melanoma (i.e., Notch, Hedgehog, and Wnt signaling).[48,49]

Notch signaling

Self-renewal via asymmetric cell division is a conserved mechanism for establishing different cell fates during development.[50] The Numb gene product, which acts as an inhibitor of both the Notch and the Hedgehog pathways, represents a segregating factor. This means that the Numb protein is only distributed into one of the two daughter cells after asymmetric cell division.[51,52] It was supposed that loss of control of asymmetric cell division might contribute to the hyperproliferation of daughter cells and thus to cancer development.[48] Although the role of Numb itself in melanoma development has not been targeted so far, the Numb-regulated Notch pathway represents an expanding field in melanoma research. It was shown that constitutive activation of Notch1 by ectopic expression of the Notch1 intracellular domain (N^{IC}) enables primary melanoma cell lines to proliferate in a serum-independent and growth factor–independent manner in vitro and to grow more aggressively with metastatic activity in vivo.[53] Interestingly, Notch signaling can be mediated by activation of the beta catenin or the mitogen-activated protein kinase/phosphatidylinositol 3-kinase-Akt pathways.[53,54] This is of particular interest because Notch signaling is implicated in the maintenance of the stem cell population in several tissue types, including several neuroecto-dermal tissues.[55] In a recent report, Moriyama and colleagues[56] highlighted the vital role of Notch activation in promoting the survival of melanocyte precursor cells, melanoblasts, and melanocyte stem cells. Using the Tyr-Cre;RBP-J$^{f/f}$ mouse model, which is characterized by the absolute absence of melanocytes in hair follicles, they detected an impaired melanogenesis secondary to disruption of the N^{IC}/RBP J–mediated transcriptional control of Notch target genes such as Hes. They could reproduce the Tyr-Cre;RBP-J$^{f/f}$ phenotype when they treated E13.5 embryos with a pharmaceutical Notch inhibitor, DAPT, before being grafted to immunodeficient mice. In vivo rescue experiments with DCT-Hes1 transgenic mice (where the Notch target Hes1 is specifically expressed in melanoblasts) in turn led to maintenance of melanocyte precursor cells. In sum, they suggested a key role of Notch signaling in maintaining the survival of embryonic murine melanoblasts and melanocyte stem cells. Furthermore, overexpression of Notch receptors and their ligands Jagged-1, Jagged-2, and Delta correlated with melanocytic tumor growth and progression in vivo.[57]

Hedgehog signaling

The Hedgehog pathway has been long known to be required for the growth of a number of human cancers, particularly sporadic nonmelanoma skin cancers,[58] and has a critical role in the growth of the dorsal brain, near the sites of origin of melanogenic precursors.[59] Stecca and colleagues[60] showed that Hedgehog SHH-GLI signaling is also active in the matrix of human hair follicles and that it is required for the normal proliferation of human melanocytes in culture. SHH-GLI signaling additionally regulated the proliferation and survival of human melanomas. The growth, recurrence, and metastasis of melanoma xenografts in mice were prevented by local or systemic interference of HH-GLI functioning. Moreover, the authors showed that oncogenic RAS-induced melanomas in transgenic mice express Gli1 and require HH-GLI signaling in vitro and in vivo.

Wnt signaling

Another important family of developmental signaling molecules in melanocyte/melanoma biology is the Wnt family. Wnt1 is involved in various cellular processes, including proliferation, migration, and differentiation in embryonic development and adult homeostasis.[61] Canonical Wnt1 signaling is critically involved in the control of the migration/invasion behavior of human mesenchymal stem cells.[62] In melanoma, several downstream targets of Wnt1, such as APC, Lef1, and beta catenin, are affected, leading to activation of this pathway. Beta catenin, the key mediator of Wnt1, can induce ubiquitous genes, such as *Myc* or *CyclinD1*, and melanocyte-specific genes, such as *Mitf-M* and *Dct*. Mitf plays a critical role in melanocyte survival, proliferation, and differentiation.[63] The role of Mitf in the maintenance of murine melanocyte stem cells in the hair follicle was elucidated by Nishimura and colleagues,[24] as reported previously. Wnt5a, which is another member of the Wnt family, signals independently from beta catenin via activation of phospholipase C, causing phospholipid turnover in the membrane, calcium release from intracellular stores, and an increase in protein kinase C (PKC) activity. Weeraratna and colleagues[64] observed an enhancement of melanoma cell invasion in direct correlation with Wnt5a expression and PKC activation. Blockage of this pathway using antibodies to Frizzled-5, the receptor for Wnt5a, led to an inhibition of PKC activity and cellular invasion. Furthermore, Wnt5a expression in human melanoma biopsies directly correlated with increasing tumor grades.[64]

THE MELANOMA STEM CELL NICHE

Stem cells reside within a specialized microenvironment called the stem cell niche. Stem cell niches are not merely repositories for stem cells, but rather, they are complex dynamic entities that actively regulate stem cell function and fate.[65,66] The composition of the niche involves heterologous cell types, matrix glycoproteins, locally and distally derived secreted signal molecules, and local metabolic conditions. Exit from the niche is usually accompanied by stem cell differentiation.[67,68]

Observations from several cancers demonstrate that the niches from normal and malignant tissue can be similar in composition. For example, glioblastoma stem cells are found in intimate contact with the aberrant tumor vasculature, mimicking the physiological niche in which neural stem cells interact with blood vessels of the subventricular zone.[67] Particularly in the case of melanoma, where the cancer stem cell niche has yet to be identified, knowledge of the physiological melanocyte stem cell niche and its key components might guide future research. Furthermore, many key players of normal and malignant stem cell niches are evolutionarily conserved and may act as additional indicators for the melanoma stem cell niche.[49] For example, tumor stroma–derived bone morphogenetic proteins (BMPs) and their antagonists are known to play a crucial role in stem cell biology as regulators of the balance between expansion and differentiation of stem cells,[49] especially in the bulge region of the hair follicle.[69,70] Using conditional gene targeting in mice, Kobielak and colleagues[69] showed that BMP receptor IA is essential for the differentiation of keratinocyte progenitor cells of the inner root sheath and hair shaft. Later on, they observed that when the *BMPR1A* gene is conditionally ablated, quiescent stem cells are activated to proliferate, causing an expansion of the niche and loss of slow-cycling cells. Surprisingly, stem cells were not lost in these experiments, but rather, they generated long-lived, tumorlike branches expressing Sox4, Lhx2, and Sonic Hedgehog and failing to terminally differentiate to make hair. As key components of *BMPR1A*-deficient stem cells, their elevated levels of both Lef1 and beta catenin were detected. Since recent data additionally suggest an involvement of BMPs in melanoma cell migration and formation of vascular networks in melanoma,[71,72] BMPs may contribute to the maintenance of the melanoma stem cell niche and, by this, may represent a possible marker to identify the melanoma stem cell niche.

Although the surrounding niche can affect the function of melanoma stem cells, this communication is not unidirectional, and there is evidence that melanoma stem cells could actively modulate their microenvironment. Using zebra fish embryos as a biosensor, Topczewska and colleagues[73] discovered that metastatic melanoma cells transplanted into the blastula-stage embryo initiated formation of either an ectopic outgrowth or a duplicate body axis.[74] The authors identified Nodal as the agent responsible for the formation of the secondary axis, as overexpression of the Nodal antagonist Lefty-1 effectively prevented its creation. In addition, Nodal knockdown induced melanoma cell differentiation, as evidenced by an increase in tyrosinase activity and decreased vascular endothelial (VE) cadherin and keratin levels.[73] As a member of the transforming growth factor (TGF) beta family, Nodal is frequently involved in processes related to embryogenesis, including the maintenance of pluripotency, regulation of mesodermal and endodermal development, regulation of neurogenesis, and proper development of axial symmetry.[75] Thus melanoma cells are able to induce the production of a biological niche, where they can prosper by transforming the microenvironment.

Expanding on the idea that melanoma cells organize their microenvironment for self-survival, further data show that melanoma cells also express several endothelium-associated genes, including VE cadherin, tyrosine kinase receptor 1, and ephrin receptor A2, which all are involved in angiogenesis.[76] According to this, melanoma cells with a high degree of differentiation plasticity can contribute to de novo formation of fluid-conducting, matrix-rich tumor blood vessels via vasculogenetic mimicry.[76]

THE ROLE OF DORMANT MELANOMA STEM CELLS IN METASTASIS AND THERAPEUTIC RESISTANCE OF MELANOMA STEM CELLS

Metastasis is a complex multistep process where cancer cells are settling at organs or tissue sites distant from the primary tumor location. Melanoma is considered to have a high malignant potency because metastatic spread arises even from very small tumor masses.[77]

The prevailing clonal selection model of metastasis contends that genetic mutations obtained late in tumorigenesis provide a selective advantage for cell subsets to metastasize.[49] Recent studies postulate that the capacity to metastasize is predetermined by genetic changes already acquired at the initial stages of tumor development.[78] Using gene expression analyses, molecular signatures could be defined that successfully predict poor prognoses of patients due to the metastatic potential of their primary tumors.[79,80] Strikingly, typical metastasis programs also seem to be activated in benign tumor lesions, such as melanocytic nevi, as reported by Gupta and colleagues.[81] Their analysis of microarray data from human nevi showed that the expression pattern of Slug, a master regulator of neural crest cell specification and migration, correlates with the expression patterns of other genes that are important for neural crest cell migration during development. Moreover, Slug is required for the metastasis of transformed melanoma cells. The authors supposed that melanocyte-specific factors present before neoplastic transformation can have a pivotal role in governing melanoma progression. However, it is also known that only distinct cancer cell subsets within primary tumors are preordained to metastasis.[82] To understand these findings in light of the cancer stem cell concept, one has to consider that stemlike cancer cells obtain both the capacity to initiate primary tumor formation and the capacity to initiate distant metastasis after escape from their niche.

A recent study correlating stemness and metastatic capacity of melanoma cells was based on the immunohistochemical expression of the stem cell markers BMI-1 and Nestin in a large panel of primary melanoma and melanoma metastases as well as in 53 melanoma cell lines.[83] Stem cell renewal factor BMI-1 is a member of the Polycomb group ring finger gene family and acts as a transcriptional repressor of the Ink4a/Arf locus encoding p16 and p14, but was also shown to increase hTERT expression.[84,85] Increased nuclear BMI-1 expression was detectable in 64% of primary melanomas, 71% of melanoma metastases, and 28% of melanoma cell lines. High Nestin expression was observed in 25% of primary melanomas, 50% of melanoma metastases, and 40% of melanoma cell lines. Interestingly, there was a significant correlation between BMI-1 and Nestin expression in cell lines and metastases indicating stem cell properties of this subset of melanoma cells. Most interestingly, a high BMI-1/low p16 expression pattern represented a significant predictor of metastasis. This suggests that BMI-1–mediated repression of p16 may contribute to an increased metastatic behavior of stem cell–like melanoma cells.

Within their niche, both normal and cancer stem cells remain quiescent for decades but can become highly dynamic once activated.[86] Dormant cancer stem cells, while not of immediate clinical concern, are also believed to be at least in part responsible for cancer recurrences because current therapies succeed at eliminating bulky disease and rapidly proliferating cells missing the quiescent tumor reservoir.[16,87] Thus the failure to eradicate most cancers, including melanoma, may be a result of misidentification of the therapeutic target.[11,86] Several reports suggest that cancer-initiating cells, particularly in melanoma, can be more resistant to chemotherapeutic drugs because of their increased expression of antiapoptotic proteins or increased activity of drug efflux mechanisms.[13,88,89] For example, Monzani and colleagues[42] proposed a phenotype for melanoma stem cells characterized by CD133 positivity and high expression of the side population marker and adenosine triphosphate (ATP) dependent drug efflux transporter (ATP-binding cassette) ABCG2. Frank and colleagues[13] demonstrated that, in malignant melanoma, chemoresistance to doxorubicin can be mediated by the drug efflux pump ABCB5.

Of all drug efflux transporters, ABCB5 attracted particular attention in the field of melanoma stem cells because ABCB5 was recently shown to be enriched in a subpopulation of melanoma cells with high tumor formation capacity in serial human-to-mouse xenotransplantation experiments.[90] In vivo genetic lineage tracking suggested that ABCB5-positive melanoma cells can also give rise to a heterogeneous progeny containing ABCB5-negative cells.[90] A further phenotypic characterization, based on the observation that ABCB5 and related ABC transporters mark physiologic progenitor cells and tumor stem cells in other tissues[91–93] revealed ABCB5-positive cells to be specifically enriched among a purified cell subset of a CD133-positive stem cell phenotype, comprising between 0.5% and 2% of all melanoma cells. Immunofluorescence double staining confirmed coexpression of both surrogate markers, CD133 and ABCB5, on 2% of G3361 melanoma cells and a distinct subset of melanoma cells in in vivo samples.[13] However, a more extensive characterization of this phenotype regarding self-renewal and transdifferentiation, especially on the single-cell level, has not yet been undertaken.[42]

Thus, in combination with other candidate markers, drug efflux transporters may represent melanoma stem cell markers and may guide future therapy strategies. Another therapeutic option would be to induce dormancy in proliferating melanoma cells to gain additional time for the patient or to release cells from quiescence to make melanoma stem cells susceptible to traditional therapies.[87]

SUMMARY AND FUTURE QUESTIONS

During the past decades, numerous reports have portrayed melanoma as an aggressive cancer with an exceptionally high degree of heterogeneity and plasticity. Today, a growing body of experimental data provides direct evidence that many characteristics of melanoma might be founded on the existence of a cell population with stem cell–like properties. Recent data from several groups suggest that this subpopulation of low-proliferating melanoma cells has the potential to self-renew and differentiate into a variety of tissue types. The term *melanoma stem cell* presumes that these cells can give rise to a heterogeneous progeny as a result of a differentiation process caricaturing normal organ development. At last, melanoma stem cells are believed to initiate and trigger primary melanoma as well as metastases.

Although most reported findings seem to be highly promising on first view, on second view, many unanswered questions exist. We do not yet know how many subpopulations of melanoma cells with stem cell properties exist. First, is there a definite number of clearly distinguishable subpopulations, or is there a continuous spectrum of cells, that is, from a master stem cell on one side of the spectrum, passing through a state of transamplifying cells that gradually lose their stemness, to differentiated tumor cells on the other side (Figure 3–2)? Second, although many cancers contain cells that display stem cell–like features, the identity of the normal cell that acquires the first genetic hit leading to the tumor-initiating cell remains elusive in melanoma. Normal cells that already have stem cell properties represent likely targets, but other mechanisms are conceivable. Human tissue possesses a much higher plasticity than has been assumed so far. As was exemplarily demonstrated by Yamanaka and colleagues,[94] human skin–derived fibroblasts can be reprogrammed into an immature pluripotent state. It could be possible that, within an established tumor, the potential of tumor cells

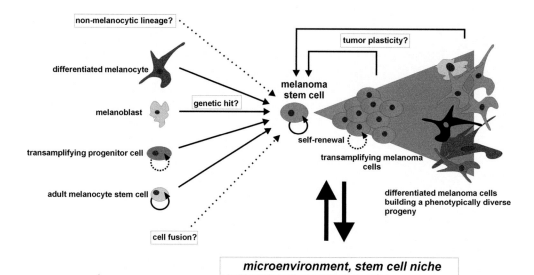

Figure 3–2: Melanoma stem cell hypothesis. Recent data indicate that there is a subpopulation of cells within melanoma that are characterized by stem cell features such as indefinite self-renewal and differentiation capacity. *Melanoma stem cell*, as an operational term, suggests that the cell of tumor origin might already be a cell with stem cell properties, like adult melanocyte stem cells or their daughter cells, called *transamplifying progenitor cells*. However, it is also conceivable that already differentiated melanocytes or their direct precursors, termed *melanoblasts*, could undergo dedifferentiation processes. Alternative models comprise cell fusion of somatic cells and blood-derived stem cells or dedifferentiation of cells from nonmelanocytic lineages such as fibroblasts. It is unclear how many subpopulations with the ability to self-renew exist in melanoma. According to normal stem cell biology, a continuous differentiation of quiescent melanoma stem cells into a heterogeneous progeny is likely (mimicked organ development). In this model, rapidly dividing transamplifying melanoma cells with a partial (dashed circle) capacity to self-renew would represent the first generation of daughter cells. Finally, the major percentage of the tumor mass would be built by differentiated melanoma cells. Although the surrounding niche can affect the function of melanoma stem cells, this communication is not unidirectional, and there is evidence that melanoma stem cells could actively modulate their microenvironments. *See color plates.*

to act as stem cells could change in response to the respective tumor's needs. Even more complicating, the question has to be addressed as to whether it is necessarily a cell from the melanocytic lineage that initiates/maintains melanoma. Could it not also be a plastic mesenchymal cell? Third, almost nothing is known so far about the niche of melanoma stem cells. What is the impact of the niche on melanoma development, maintenance, and metastasis? And fourth, how can we detect and study melanoma stem cells in the future? The first upcoming markers, such as CD20, CD133, or ABC transporters, are promising, but all have weaknesses in reproducibility, consistency, and the power to predict stemness, especially when applied to different melanoma cell lines or in vivo samples. In the future, better markers and, most likely, marker combinations are needed for a reliable characterization of melanoma stem cell subpopulations. As a long-term perspective, melanoma stem cell research will certainly influence and, it is hoped, improve the diagnosis, prognosis, and therapy of melanoma. Traditional treatments might be recalibrated and investigational therapies developed focusing on the ability to target the melanoma stem cell subpopulation and its specific biochemical pathways.

REFERENCES

1. DeVita, V., Hellman, S., Rosenberg, S.A. (2005). *Cancer: Principles and Practice of Oncology*. Philadelphia: Lippincott Williams and Wilkins.
2. Balch, C.M., Buzaid, A.C., Soong, S.J., Atkins, M.B., Cascinelli, N., Coit, D.G., Fleming, I.D., Gershenwald, J.E., Houghton, A., Jr., Kirkwood, J.M., McMasters, K.M., Mihm, M.F., Morton, D.L., Reintgen, D.S., Ross, M.I., Sober, A., Thompson, J.A., and Thompson, J.F. (2001). Final version of the American Joint Committee on Cancer staging system for cutaneous melanoma. *J Clin Oncol* **19**, 3635–3648.
3. Clark, W.H., Jr., Elder, D.E., Guerry, D.T., Epstein, M.N., Greene, M.H., and Van Horn, M. (1984). A study of tumor progression: the precursor lesions of superficial spreading and nodular melanoma. *Hum Pathol* **15**, 1147–1165.
4. Herlyn, M., Thurin, J., Balaban, G., Bennicelli, J.L., Herlyn, D., Elder, D.E., Bondi, E., Guerry, D., Nowell, P., Clark, W.H., and Koprowski H. (1985). Characteristics of cultured human melanocytes isolated from different stages of tumor progression. *Cancer Res* **45**, 5670–5676.
5. Hussein, M.R. (2004). Genetic pathways to melanoma tumorigenesis. *J Clin Pathol* **57**, 797–801.
6. Lucas, C.R., Sanders, L.L., Murray, J.C., Myers, S.A., Hall, R.P., and Grichnik, J.M. (2003). Early melanoma detection: nonuniform dermoscopic features and growth. *J Am Acad Dermatol* **48**, 663–671.
7. Roesch, A., Burgdorf, W., Stolz, W., Landthaler, M., and Vogt, T. (2006). Dermatoscopy of "dysplastic nevi": a beacon in diagnostic darkness. *Eur J Dermatol* **16**, 479–493.
8. Kath, R., Jambrosic, J.A., Holland, L., Rodeck, U., and Herlyn, M. (1991). Development of invasive and growth factor-independent cell variants from primary human melanomas. *Cancer Res* **51**, 2205–2211.
9. Rasheed, S., Mao, Z., Chan, J.M., and Chan, L.S. (2005). Is melanoma a stem cell tumor? Identification of neurogenic proteins in trans-differentiated cells. *J Transl Med* **3**, 14.
10. Buac, K., and Pavan, W.J. (2007). Stem cells of the melanocyte lineage. *Cancer Biomark* **3**, 203–209.
11. Reya, T., Morrison, S.J., Clarke, M.F., and Weissman, I.L. (2001). Stem cells, cancer, and cancer stem cells. *Nature* **414**, 105–111.
12. Fang, D., Nguyen, T.K., Leishear, K., Finko, R., Kulp, A.N., Hotz, S., Van Belle, P.A., Xu, X., Elder, D.E., and Herlyn, M. (2005). A tumorigenic subpopulation with stem cell properties in melanomas. *Cancer Res* **65**, 9328–9337.
13. Frank, N.Y., Margaryan, A., Huang, Y., Schatton, T., Waaga-Gasser, A.M., Gasser, M., Sayegh, M.H., Sadee, W., and Frank, M.H. (2005). ABCB5-mediated doxorubicin transport and chemoresistance in human malignant melanoma. *Cancer Res* **65**, 4320–4333.
14. Grichnik, J.M., Burch, J.A., Schulteis, R.D., Shan, S., Liu, J., Darrow, T.L., Vervaert, C.E., and Seigler, H.F. (2006). Melanoma, a tumor based on a mutant stem cell? *J Invest Dermatol* **126**, 142–153.
15. Tan, B.T., Park, C.Y., Ailles, L.E., and Weissman, I.L. (2006). The cancer stem cell hypothesis: a work in progress. *Lab Invest* **86**, 1203–1207.
16. Dalerba, P., Cho, R.W., and Clarke, M.F. (2007). Cancer stem cells: models and concepts. *Annu Rev Med* **58**, 267–284.
17. Mayer, T.C. (1973). The migratory pathway of neural crest cells into the skin of mouse embryos. *Dev Biol* **34**, 39–46.
18. Wehrle-Haller, B., and Weston, J.A. (1995). Soluble and cell-bound forms of steel factor activity play distinct roles in melanocyte precursor dispersal and survival on the lateral neural crest migration pathway. *Development* **121**, 731–742.

19. Cramer, S.F. (1984). The neoplastic development of malignant melanoma: a biological rationale. *Am J Dermatopathol* **6**(Suppl), 299–308.
20. Steingrimsson, E., Copeland, N.G., and Jenkins, N.A. (2005). Melanocyte stem cell maintenance and hair graying. *Cell* **121**, 9–12.
21. Nishimura, E.K., Jordan, S.A., Oshima, H., Yoshida, H., Osawa, M., Moriyama, M., Jackson, I.J., Barrandon, Y., Miyachi, Y., and Nishikawa, S. (2002). Dominant role of the niche in melanocyte stem-cell fate determination. *Nature* **416**, 854–860.
22. Osawa, M., Egawa, G., Mak, S.S., Moriyama, M., Freter, R., Yonetani, S., Beermann, F., and Nishikawa, S. (2005). Molecular characterization of melanocyte stem cells in their niche. *Development* **132**, 5589–5599.
23. Yu, H., Fang, D., Kumar, S.M., Li, L., Nguyen, T.K., Acs, G., Herlyn, M., and Xu, X. (2006). Isolation of a novel population of multipotent adult stem cells from human hair follicles. *Am J Pathol* **168**, 1879–1888.
24. Nishimura, E.K., Granter, S.R., and Fisher, D.E. (2005). Mechanisms of hair graying: incomplete melanocyte stem cell maintenance in the niche. *Science* **307**, 720–724.
25. Toma, J.G., McKenzie, I.A., Bagli, D., and Miller, F.D. (2005). Isolation and characterization of multipotent skin-derived precursors from human skin. *Stem Cells* **23**, 727–737.
26. Crigler, L., Kazhanie, A., Yoon, T.J., Zakhari, J., Anders, J., Taylor, B., and Virador, V.M. (2007). Isolation of a mesenchymal cell population from murine dermis that contains progenitors of multiple cell lineages. *Faseb J* **21**, 2050–2063.
27. Yu, H.S. (2002). Melanocyte destruction and repigmentation in vitiligo: a model for nerve cell damage and regrowth. *J Biomed Sci* **9**, 564–573.
28. Hendrix, M.S.E., Seftor, E.A., Meltzer, P.S., Hess, A.R., Gruman, L.M., Nickoloff, B.J., Miele, L., Sheriff, D.D., Schatteman, G.C., Bourdon, M.A., and Sefotro, R.E.B. (2003). The stem cell plasticity of aggressive melanoma tumor cells. In *Stem Cells Handbook*, pp. 297–306 (Sell, S., ed). Totowa, NJ: Humana Press.
29. Hendrix, M.J., Seftor, E.A., Hess, A.R., and Seftor, R.E. (2003). Molecular plasticity of human melanoma cells. *Oncogene* **22**, 3070–3075.
30. Wiese, C., Rolletschek, A., Kania, G., Blyszczuk, P., Tarasov, K.V., Tarasova, Y., Wersto, R.P., Boheler, K.R., and Wobus, A.M. (2004). Nestin expression – a property of multi-lineage progenitor cells? *Cell Mol Life Sci* **61**, 2510–2522.
31. Brocker, E.B., Magiera, H., and Herlyn, M. (1991). Nerve growth and expression of receptors for nerve growth factor in tumors of melanocyte origin. *J Invest Dermatol* **96**, 662–665.
32. Weiss, S., Reynolds, B.A., Vescovi, A.L., Morshead, C., Craig, C.G., and Van Der Kooy, D. (1996). Is there a neural stem cell in the mammalian forebrain? *Trends Neurosci* **19**, 387–393.
33. Toma, J.G., Akhavan, M., Fernandes, K.J., Barnabe-Heider, F., Sadikot, A., Kaplan, D.R., and Miller, F.D. (2001). Isolation of multipotent adult stem cells from the dermis of mammalian skin. *Nat Cell Biol* **3**, 778–784.
34. Dontu, G., Abdallah, W.M., Foley, J.M., Jackson, K.W., Clarke, M.F., Kawamura, M.J., and Wicha, M.S. (2003). In vitro propagation and transcriptional profiling of human mammary stem/progenitor cells. *Genes Dev* **17**, 1253–1270.
35. Singh, S.K., Clarke, I.D., Terasaki, M., Bonn, V.E., Hawkins, C., Squire, J., and Dirks, P.B. (2003). Identification of a cancer stem cell in human brain tumors. *Cancer Res* **63**, 5821–5828.
36. Singh, S.K., Hawkins, C., Clarke, I.D., Squire, J.A., Bayani, J., Hide, T., Henkelman, R.M., Cusimano, M.D., and Dirks, P.B. (2004). Identification of human brain tumour initiating cells. *Nature* **432**, 396–401.
37. Bittner, M., Meltzer, P., Chen, Y., Jiang, Y., Seftor, E., Hendrix, M., Radmacher, M., Simon, R., Yakhini, Z., Ben-Dor, A., Sampas, N., Dougherty, E., Wang, E., Marincola, F., Gooden, C., Lueders, J., Glatfelter, A., Pollock, P., Carpten, J., Gillanders, E., Leja,

D., Dietrich, K., Beaudry, C., Berens, M., Alberts, D., and Sondak, V. (2000). Molecular classification of cutaneous malignant melanoma by gene expression profiling. *Nature* **406**, 536–540.

38. Coiffier, B., Lepage, E., Briere, J., Herbrecht, R., Tilly, H., Bouabdallah, R., Morel, P., Van Den Neste, E., Salles, G., Gaulard, P., Reyes, F., Lederlin, P., and Gisselbrecht, C. (2002). CHOP chemotherapy plus rituximab compared with CHOP alone in elderly patients with diffuse large-B-cell lymphoma. *N Engl J Med* **346**, 235–242.

39. Salmaggi, A., Boiardi, A., Gelati, M., Russo, A., Calatozzolo, C., Ciusani, E., Sciacca, F.L., Ottolina, A., Parati, E.A., La Porta, C., Alessandri, G., Marras, C., Croci, D., and De Rossi, M. (2006). Glioblastoma-derived tumorospheres identify a population of tumor stem-like cells with angiogenic potential and enhanced multidrug resistance phenotype. *Glia* **54**, 850–860.

40. Neuzil, J., Stantic, M., Zobalova, R., Chladova, J., Wang, X., Prochazka, L., Dong, L., Andera, L., and Ralph, S.J. (2007). Tumour-initiating cells vs. cancer "stem" cells and CD133: what's in the name? *Biochem Biophys Res Commun* **355**, 855–859.

41. Klein, W.M., Wu, B.P., Zhao, S., Wu, H., Klein-Szanto, A.J., and Tahan, S.R. (2007). Increased expression of stem cell markers in malignant melanoma. *Mod Pathol* **20**, 102–107.

42. Monzani, E., Facchetti, F., Galmozzi, E., Corsini, E., Benetti, A., Cavazzin, C., Gritti, A., Piccinini, A., Porro, D., Santinami, M., Invernici, G., Parati, E., Alessandri, G., and La Porta, C.A. (2007). Melanoma contains CD133 and ABCG2 positive cells with enhanced tumourigenic potential. *Eur J Cancer* **43**, 935–946.

43. Goodell, M.A., Brose, K., Paradis, G., Conner, A.S., and Mulligan, R.C. (1996). Isolation and functional properties of murine hematopoietic stem cells that are replicating in vivo. *J Exp Med* **183**, 1797–1806.

44. Wang, E., Voiculescu, S., Le Poole, I.C., El-Gamil, M., Li, X., Sabatino, M., Robbins, P.F., Nickoloff, B.J., and Marincola, F.M. (2006). Clonal persistence and evolution during a decade of recurrent melanoma. *J Invest Dermatol* **126**, 1372–1377.

45. Wiltshire, R.N., Duray, P., Bittner, M.L., Visakorpi, T., Meltzer, P.S., Tuthill, R.J., Liotta, L.A., and Trent, J.M. (1995). Direct visualization of the clonal progression of primary cutaneous melanoma: application of tissue microdissection and comparative genomic hybridization. *Cancer Res* **55**, 3954–3957.

46. Bastian, B.C., LeBoit, P.E., Hamm, H., Brocker, E.B., and Pinkel, D. (1998). Chromosomal gains and losses in primary cutaneous melanomas detected by comparative genomic hybridization. *Cancer Res* **58**, 2170–2175.

47. Wiltshire, R.N., Dennis, T.R., Sondak, V.K., Meltzer, P.S., and Trent, J.M. (2001). Application of molecular cytogenetic techniques in a case study of human cutaneous metastatic melanoma. *Cancer Genet Cytogenet* **131**, 97–103.

48. Tysnes, B.B., and Bjerkvig, R. (2007). Cancer initiation and progression: involvement of stem cells and the microenvironment. *Biochim Biophys Acta* **1775**, 283–297.

49. Li, F., Tiede, B., Massague, J., and Kang, Y. (2007). Beyond tumorigenesis: cancer stem cells in metastasis. *Cell Res* **17**, 3–14.

50. Horvitz, H.R., and Herskowitz, I. (1992). Mechanisms of asymmetric cell division: two Bs or not two Bs, that is the question. *Cell* **68**, 237–255.

51. Roegiers, F., and Jan, Y.N. (2004). Asymmetric cell division. *Curr Opin Cell Biol* **16**, 195–205.

52. Di Marcotullio, L., Ferretti, E., Greco, A., De Smaele, E., Po, A., Sico, M.A., Alimandi, M., Giannini, G., Maroder, M., Screpanti, I., and Gulino, A. (2006). Numb is a suppressor of Hedgehog signalling and targets Gli1 for Itch-dependent ubiquitination. *Nat Cell Biol* **8**, 1415–1423.

53. Liu, Z.J., Xiao, M., Balint, K., Smalley, K.S., Brafford, P., Qiu, R., Pinnix, C.C., Li, X., and Herlyn, M. (2006). Notch1 signaling promotes primary melanoma progression by activating mitogen-activated protein kinase/phosphatidylinositol 3-kinase-Akt pathways and up-regulating N-cadherin expression. *Cancer Res* **66**, 4182–4190.

54. Balint, K., Xiao, M., Pinnix, C.C., Soma, A., Veres, I., Juhasz, I., Brown, E.J., Capobianco, A.J., Herlyn, M., and Liu, Z.J. (2005). Activation of Notch1 signaling is required for beta-catenin-mediated human primary melanoma progression. *J Clin Invest* **115**, 3166–3176.

55. Pinnix, C.C., and Herlyn, M. (2007). The many faces of Notch signaling in skin-derived cells. *Pigment Cell Res* **20**, 458–465.

56. Moriyama, M., Osawa, M., Mak, S.S., Ohtsuka, T., Yamamoto, N., Han, H., Delmas, V., Kageyama, R., Beermann, F., Larue, L., and Nishikawa, S. (2006). Notch signaling via Hes1 transcription factor maintains survival of melanoblasts and melanocyte stem cells. *J Cell Biol* **173**, 333–339.

57. Massi, D., Tarantini, F., Franchi, A., Paglierani, M., Di Serio, C., Pellerito, S., Leoncini, G., Cirino, G., Geppetti, P., and Santucci, M. (2006). Evidence for differential expression of Notch receptors and their ligands in melanocytic nevi and cutaneous malignant melanoma. *Mod Pathol* **19**, 246–254.

58. Dahmane, N., Lee, J., Robins, P., Heller, P., and Ruiz i Altaba, A. (1997). Activation of the transcription factor Gli1 and the Sonic hedgehog signalling pathway in skin tumours. *Nature* **389**, 876–881.

59. Le Douarin, N.M., and Dupin, E. (2003). Multipotentiality of the neural crest. *Curr Opin Genet Dev* **13**, 529–536.

60. Stecca, B., Mas, C., Clement, V., Zbinden, M., Correa, R., Piguet, V., Beermann, F., and Ruiz, I.A.A. (2007). Melanomas require HEDGEHOG-GLI signaling regulated by interactions between GLI1 and the RAS-MEK/AKT pathways. *Proc Natl Acad Sci U S A* **104**, 5895–5900.

61. Larue, L., and Delmas, V. (2006). The WNT/Beta-catenin pathway in melanoma. *Front Biosci* **11**, 733–742.

62. Neth, P., Ries, C., Karow, M., Egea, V., Ilmer, M., and Jochum, M. (2007). The Wnt signal transduction pathway in stem cells and cancer cells: influence on cellular invasion. *Stem Cell Rev* **3**, 18–29.

63. Miller, A.J., and Mihm, M.C., Jr. (2006). Melanoma. *N Engl J Med* **355**, 51–65.

64. Weeraratna, A.T., Jiang, Y., Hostetter, G., Rosenblatt, K., Duray, P., Bittner, M., and Trent, J.M. (2002). Wnt5a signaling directly affects cell motility and invasion of metastatic melanoma. *Cancer Cell* **1**, 279–288.

65. Spradling, A., Drummond-Barbosa, D., and Kai, T. (2001). Stem cells find their niche. *Nature* **414**, 98–104.

66. Scadden, D.T. (2006). The stem-cell niche as an entity of action. *Nature* **441**, 1075–1079.

67. Gilbertson, R.J., and Rich, J.N. (2007). Making a tumour's bed: glioblastoma stem cells and the vascular niche. *Nat Rev Cancer* **7**, 733–736.

68. Mitsiadis, T.A., Barrandon, O., Rochat, A., Barrandon, Y., and De Bari, C. (2007). Stem cell niches in mammals. *Exp Cell Res* **313**, 3377–3385.

69. Kobielak, K., Pasolli, H.A., Alonso, L., Polak, L., and Fuchs, E. (2003). Defining BMP functions in the hair follicle by conditional ablation of BMP receptor IA. *J Cell Biol* **163**, 609–623.

70. Kobielak, K., Stokes, N., de la Cruz, J., Polak, L., and Fuchs, E. (2007). Loss of a quiescent niche but not follicle stem cells in the absence of bone morphogenetic protein signaling. *Proc Natl Acad Sci U S A* **104**, 10063–10068.

71. Rothhammer, T., Bataille, F., Spruss, T., Eissner, G., and Bosserhoff, A.K. (2007). Functional implication of BMP4 expression on angiogenesis in malignant melanoma. *Oncogene* **26**, 4158–4170.

72. Rothhammer, T., Poser, I., Soncin, F., Bataille, F., Moser, M., and Bosserhoff, A.K. (2005). Bone morphogenic proteins are overexpressed in malignant melanoma and promote cell invasion and migration. *Cancer Res* **65**, 448–456.

73. Topczewska, J.M., Postovit, L.M., Margaryan, N.V., Sam, A., Hess, A.R., Wheaton, W.W., Nickoloff, B.J., Topczewski, J., and Hendrix, M.J. (2006). Embryonic and

tumorigenic pathways converge via Nodal signaling: role in melanoma aggressiveness. *Nat Med* **12**, 925–932.

74. Hendrix, M.J., Seftor, E.A., Seftor, R.E., Kasemeier-Kulesa, J., Kulesa, P.M., and Postovit, L.M. (2007). Reprogramming metastatic tumour cells with embryonic microenvironments. *Nat Rev Cancer* **7**, 246–255.

75. Lee, J.T., and Herlyn, M. (2006). Embryogenesis meets tumorigenesis. *Nat Med* **12**, 882–884.

76. Hendrix, M.J., Seftor, E.A., Hess, A.R., and Seftor, R.E. (2003). Vasculogenic mimicry and tumour-cell plasticity: lessons from melanoma. *Nat Rev Cancer* **3**, 411–421.

77. Meier, F., Will, S., Ellwanger, U., Schlagenhauff, B., Schittek, B., Rassner, G., and Garbe, C. (2002). Metastatic pathways and time courses in the orderly progression of cutaneous melanoma. *Br J Dermatol* **147**, 62–70.

78. Bernards, R., and Weinberg, R.A. (2002). A progression puzzle. *Nature* **418**, 823.

79. van't Veer, L.J., Dai, H., van de Vijver, M.J., He, Y.D., Hart, A.A., Mao, M., Peterse, H.L., van der Kooy, K., Marton, M.J., Witteveen, A.T., Schreiber, G.J., Kerkhoven, R.M., Roberts, C., Linsley, P.S., Bernards, R., and Friend, S.H. (2002). Gene expression profiling predicts clinical outcome of breast cancer. *Nature* **415**, 530–536.

80. Ramaswamy, S., Ross, K.N., Lander, E.S., and Golub, T.R. (2003). A molecular signature of metastasis in primary solid tumors. *Nat Genet* **33**, 49–54.

81. Gupta, P.B., Kuperwasser, C., Brunet, J.P., Ramaswamy, S., Kuo, W.L., Gray, J.W., Naber, S.P., and Weinberg, R.A. (2005). The melanocyte differentiation program predisposes to metastasis after neoplastic transformation. *Nat Genet* **37**, 1047–1054.

82. Minn, A.J., Gupta, G.P., Siegel, P.M., Bos, P.D., Shu, W., Giri, D.D., Viale, A., Olshen, A.B., Gerald, W.L., and Massague, J. (2005). Genes that mediate breast cancer metastasis to lung. *Nature* **436**, 518–524.

83. Mihic-Probst, D., Kuster, A., Kilgus, S., Bode-Lesniewska, B., Ingold-Heppner, B., Leung, C., Storz, M., Seifert, B., Marino, S., Schraml, P., Dummer, R., and Moch, H. (2007). Consistent expression of the stem cell renewal factor BMI-1 in primary and metastatic melanoma. *Int J Cancer* **121**, 1764–1770.

84. Lessard, J., Baban, S., and Sauvageau, G. (1998). Stage-specific expression of polycomb group genes in human bone marrow cells. *Blood* **91**, 1216–1224.

85. Dimri, G.P., Martinez, J.L., Jacobs, J.J., Keblusek, P., Itahana, K., Van Lohuizen, M., Campisi, J., Wazer, D.E., and Band, V. (2002). The Bmi-1 oncogene induces telomerase activity and immortalizes human mammary epithelial cells. *Cancer Res* **62**, 4736–4745.

86. Bissell, M.J., and Labarge, M.A. (2005). Context, tissue plasticity, and cancer: are tumor stem cells also regulated by the microenvironment? *Cancer Cell* **7**, 17–23.

87. Townson, J.L., and Chambers, A.F. (2006). Dormancy of solitary metastatic cells. *Cell Cycle* **5**, 1744–1750.

88. Doyle, L.A., and Ross, D.D. (2003). Multidrug resistance mediated by the breast cancer resistance protein BCRP (ABCG2). *Oncogene* **22**, 7340–7358.

89. Liu, G., Yuan, X., Zeng, Z., Tunici, P., Ng, H., Abdulkadir, I.R., Lu, L., Irvin, D., Black, K.L., and Yu, J.S. (2006). Analysis of gene expression and chemoresistance of CD133+ cancer stem cells in glioblastoma. *Mol Cancer* **5**, 67.

90. Schatton, T., Murphy, G.F., Frank, N.Y., Yamaura, K., Waaga-Gasser, A.M., Gasser, M., Zhan, Q., Jordan, S., Duncan, L.M., Weishaupt, C., Fuhlbrigge, R.C., Kupper, T.S., Sayegh, M.H., and Frank, M.H. (2008). Identification of cells initiating human melanomas. *Nature* **451**, 345–349.

91. Chaudhary, P.M., and Roninson, I.B. (1991). Expression and activity of P-glycoprotein, a multidrug efflux pump, in human hematopoietic stem cells. *Cell* **66**, 85–94.

92. Wulf, G.G., Wang, R.Y., Kuehnle, I., Weidner, D., Marini, F., Brenner, M.K., Andreeff, M., and Goodell, M.A. (2001). A leukemic stem cell with intrinsic drug efflux capacity in acute myeloid leukemia. *Blood* **98**, 1166–1173.

93. Zhou, S., Schuetz, J.D., Bunting, K.D., Colapietro, A.M., Sampath, J., Morris, J.J., Lagutina, I., Grosveld, G.C., Osawa, M., Nakauchi, H., and Sorrentino, B.P. (2001). The ABC transporter Bcrp1/ABCG2 is expressed in a wide variety of stem cells and is a molecular determinant of the side-population phenotype. *Nat Med* **7**, 1028–1034.

94. Takahashi, K., Tanabe, K., Ohnuki, M., Narita, M., Ichisaka, T., Tomoda, K., and Yamanaka, S. (2007). Induction of pluripotent stem cells from adult human fibroblasts by defined factors. *Cell* **131**, 861–872.

4 Mammospheres and breast carcinoma

Massimiliano Bonafe

University of Bologna, Bologna, Italy

In this chapter, recent literature on human and mouse normal mammary gland and breast cancer stem/progenitor cells will be reviewed. Part of these data will be gathered from studies performed on stem/progenitor cells in vitro expanded as multicellular spheroids, called *mammospheres* (MS). It will be highlighted that the available data support the notion that normal and putative cancer stem cell gene expression patterns are close to the basal-like phenotype. Such a phenotype identifies the normal mammary gland cell compartment that is supposed to harbor stem/progenitor cells as well as a subset of highly aggressive/metastatic breast carcinomas lacking estrogen receptor alpha (ERα) expression and overexpressing cytokeratin 5/6 (CK5/6), epidermal growth factor receptor (EGFr), interleukin-6 (IL-6), Notch-3, Jagged-1, carbonic anhydrase isoenzyme-IX (CA-IX), vimentin, and SNAI gene family members. It will be then argued that understanding the regulation of basal-like gene expression profile is expected to provide an insight on normal and cancer stem cell–controlling mechanisms. In particular, it will be pointed out that the pro-inflammatory cytokine IL-6, the stem cell regulatory gene Notch-3, its ligand Jagged-1, and the hypoxia survival gene CA-IX share a molecular machinery that promotes invasive behavior and survival of normal mammary gland and breast cancer stem cells. It will be also reported that, in normal mammary gland and breast cancer cells, the SNAI gene family members govern the onset of an undifferentiated mesenchymal phenotype that parallels the gain of a basal-like/stem cell–like phenotype. It will then be explained that IL-6 and SNAI2 gene expression in breast cancer stem cells can be regarded as the signature of an inflammatory/hypoxic response pathway that is overexpressed in the stem cell–like lymphovascular emboli of inflammatory breast carcinomas. It will be concluded that the targeting of inflammatory/hypoxic pathways is a feasible approach to undertake a breast cancer stem cell–specific therapy that may also be applied to highly aggressive breast carcinoma subtypes in which stem cell–like features are peculiarly overt, that is, the basal-like and inflammatory breast carcinomas.

THE BASAL-LIKE PHENOTYPE IN NORMAL MAMMARY GLAND STEM CELLS

The presence of stem cells in the normal mammary gland has long been postulated on the basis of its regenerative ability during puberty and reproductive cycles, the high cell turnover, and the capability of mammary gland fragments to reconstitute a functional gland on serial transplant.[1] In past years, it was shown that a single cell with a definite membrane phenotype can reconstitute the entire mouse mammary gland.[2-4] The analysis of X chromosome inactivation supported the notion that the fundamental units of the human mammary gland, the terminal duct lobular units, have a monoclonal (stem cell) origin.[5,6] In line with the notion that mammary gland stem/progenitor cells are a minor fraction of the basal cell compartment,[7-11] isolated normal mammary gland stem cells have been found to express a basal-like phenotype.[3,4,12-15] Basal-like phenotype includes the expression of CK5/6 and EGFr and the absence of ERα expression.[16-20] In humans, the in vitro model of nonadherent multicellular spheroids, called MS,allowed to propagate from the normal mammary gland a cell population containing a large proportion of cells endowed with a stem/progenitor cell phenotype and showing upregulation of a basal-like gene expression pattern.[21-28] The tight association between the basal-like and stem cell–like phenotypes gained recent importance, when a breast cancer subtype, the basal-like carcinoma, was shown to express a stem cell–like gene expression profile.[26,27,29-32] In support of the link between mammary gland stem cells and the basal-like phenotype, it was also observed that breast tumors that arose in BRCA-I mutation carrier women and in BRCA-1 defective mice disclosed a basal-like phenotype in almost all cases,[33,34] and that the in vitro knockdown of the BRCA-I gene leads to an increase in the expansion of MS.[25] Overall, it can be proposed that normal stem/progenitor cells of the mammary gland disclose a basal-like gene expression profile.

THE BASAL-LIKE PHENOTYPE OF BREAST CANCER STEM CELLS

The stem cell phenotype in breast cancer cells was demonstrated by showing that the bulk of tumor cells contains a minor, phenotypically identifiable cell population that can establish transplantable xenografts, recapitulating the heterogeneous features of the original tumor mass.[35–37] Such a cell population, named cancer stem cells or tumor-initiating cells, can be isolated by the immunosorting of breast cancer cells that express the hyaluronian receptor CD44 (CD44+), a gene that is overexpressed in basal-like tumors[38] and lack the expression of CD24, an endogenous inhibitor of the chemochine receptor CXCR4.[35,39] CD44+ cells isolated from ductal breast carcinoma and from normal mammary gland were found to express low levels of ERα and high levels of CK5[40] (Table 4–1). Accordingly, a CD44+ and CK5- and ERα-negative cell subpopulation was identified in ductal breast carcinoma xenografts and disclosed the capability to generate CK5-negative/ERα-positive cells and to reestablish xenografts.[54] In keeping with these findings, we observed that, irrespective of the tumor phenotype, MS from ductal breast carcinoma lack ERα expression and overexpress CD44, CK5, and EGFr[26–28] (Table 4–1). These data allow us to conclude that similarly to their normal counterparts, breast cancer stem cells disclose a basal-like phenotype. Moreover, because the multicellular spheroids of luminal breast cancer cells (e.g., MCF-7) show a basal-like profile and an enhanced tumor-forming capacity[26–28,52,53] (P. Sansone 2009, manuscript in preparation) (Table 4–1), and the amount of cells showing a basal-like phenotype correlates with breast cancer cell line aggressiveness,[55,56] it can be proposed that the highly aggressive behavior of basal-like tumors may be the consequence of their close relationship to the stem cell phenotype.

GENES ASSOCIATED WITH THE BASAL-LIKE PHENOTYPE: UNDERSTANDING THE BREAST CANCER STEM CELL PHENOTYPE REGULATION IN MAMMOSPHERES

In keeping with the preceding reasoning, it can be observed that some genes overexpressed in basal-like tumor are upregulated in models of normal mammary gland and breast cancer stem cells, including normal and tumor MS (Table 4–1). Among these genes, vimentin, a mesenchymal hallmark, and SNAI gene family members, which promote the onset of a mesenchymal gene expression profile and phenotype in epithelial cells, can be enrolled.[51,57,58] These data demonstrate that a tight connection exists among the basal, the stem, and the mesenchymal phenotypes in normal and tumor mammary gland stem/progenitor cells.[21,28,39,47,50,51,58–63] Indeed, such an overlapping is represented in MS (Figure 4–1). In the following two paragraphs, the functional relationship among the genes belonging to the basal-like phenotype will be highlighted, starting from recent studies on MS.[26–28] In particular, an lL-6–controlled gene network will

Table 4–1. Expression of basal-like/stem cell–like phenotype in different normal and breast cancer stem cell models

	Basal like tumors vs. other subtypes[a]	Isolated stem cells vs. other[b]	Normal MS vs. adherent[c]	Tumor MS vs. tumor mass[d]	CD44+ vs. CD44[e]	Spheroids vs. adherent[f]
CK5	+	+	+	+	+	+
EGFr	+	+	+	+	N/A	+
ERα	−	−	−	−	−	−
Notch-3	+	+	+	+	+	+
Jagged-1	+	+	+	+	+	+
IL-6	+	N/A	+	+	+	+
CD44	+	+	+	+	+	+
CD24	−	−	−	−	−	−
CA-IX	+	N/A	+	+	N/A	+
SNAI2	+	N/A	+	+	+	+
Vimentin	+	+	+	+	+	+

[a] Charafe-Jouffret et al.[39]; Bertucci et al.[41] Reedjik et al.[42]; Reedjik et al.[43]; Dickson et al.[44]; Sansome et al.[26,27]; Storci et al.[28]; Brennan et al.[45]; Kuroda et al.[46]; Sarrió et al.[47]; Nielsen et al.[16]; Fulford et al.[17]; Rakha et al.[48]; Livasy et al.[19]; Reis-Filho et al.[20]
[b] Shackelton et al.[3]; Stingl et al.[4]; Sleeman et al.[49]; Vaillant et al.[13]; Asselin Labat et al.[14]; Regan and Smalley.[15]
[c] Sansone et al.[26,27]; Storci et al.[28]; Liao et al.[50]
[d] Sansone et al.[26,27]; Storci et al.[28]; Dontu et al.[21,22]
[e] Shipitsin et al.[40]; Mani et al.[51]
[f] Sansone et al.[26,27]; Storci et al.[28]; Phillips et al.[52]; Ponti et al.[53]; P. Sansone and G. Storci, manuscript in preparation, 2009.

be proposed to be active in the promotion of survival and invasive behavior of normal mammary gland and breast cancer stem cells. Moreover, a crucial role of the SNAI2-driven mechanism in the control of stem cell differentiation and the gain of a basal/stem cell–like invasive phenotype will be pinpointed. It will then be proposed that both IL-6– and SNAI2-controlled pathways may share in a hypoxia/inflammatory response pathway that may regulate normal stem cells and, when deranged, promote the stem cell phenotype in breast cancer cells.

INTERLEUKIN-6: A PREDICTABLE LINK BETWEEN INFLAMMATION AND STEM CELL REGULATION

Elevated levels of the pro-inflammatory cytokine IL-6 are predictors of poor prognosis in breast cancer patients.[64] IL-6 is expressed in vitro by aggressive mesenchymal/basal-like breast cancer cell lines and tissues and promotes malignant features in breast cancer cells and xenografts.[39,56,61,64–67] MS were found to express IL-6 and gp130, the common receptor subunit for IL-6 family cytokines.[21,26] We recently found that IL-6 is overexpressed in MS from invasive ductal breast carcinoma.[26] We then reported that, in breast cancer cells and MS, exogenous and autocrine IL-6 upregulates the expression of three genes associated with the basal-like phenotype, that is, Notch-3, Jagged-1, and CA-IX.[26–28,42–45,68,69] Moreover, we observed that Notch-3 controls Jagged-1 and CA-IX gene expression via the activation of the extracellular receptor kinase

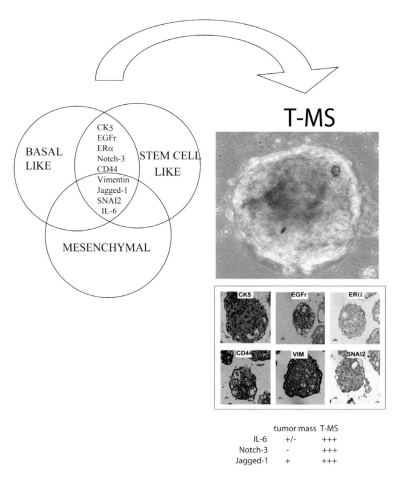

Figure 4–1: Schematic of the overlapping among the basal like, the stem cell, and the mesenchymal phenotype and its representation in human ductal breast carcinoma–derived (T–) MS.[26–28] A similar phenotype has been observed in normal MS[21–28] and murine MS.[50] *See color plates.*

(ERK)[26,27] (Figure 4–2). We also found that the Notch-3 gene is a target of Stat-3, the major intracellular mediator of IL-6 receptor signaling,[70,71] and that the Notch-3–dependent activation of ERK requires the activity of the EGFr receptor and the Ras pathway (Figure 4–2). Independent observation revealed that both EGFr and Ras activation are potent activators of IL-6 gene expression in breast cancer cells, thus promoting a potential autocrine loop.[67,70] Interestingly, features of EGFr/Ras/ERK axis activation were observed in basal tumors,[72] in the invasive signature of CD44+ breast cancer stem cells,[29] and Ras/ERK activation was found to elicit a mesenchymal/CD44+ stem cell phenotype in mammary gland epithelial cells.[58] Collectively, the gene network described reveals a functional link among several genes belonging to the basal-like phenotype. Features of such a functional relationship between gp130/Stat-3 signaling and the EGFr pathway have been previously reported in primary mammary gland epithelial cells and neural progenitors.[73,74] In this regard, a gp130/Stat-3–dependent upregulation of Notch gene expression was described in neurospheres[75] that, similarly to MS, grow in an EGF-supplemented serum-free medium and require functional

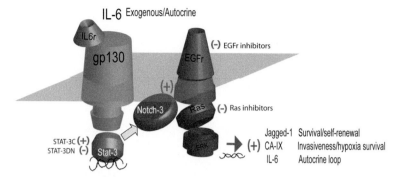

Figure 4–2: Schematic of IL-6–controlled activities in MS and breast cancer cells. IL-6 pathway activation promotes an aggressive/stem cell–like phenotype in breast cancer cells and triggers the onset of an autocrine IL-6 loop.[26,27] EGFr receptor and Ras specific inhibitors block the pathway, and the transfection of constitutively active (STAT-3C) or dominant negative (STAT3-DN) Stat-3 proteins modulates Notch-3 gene expression (P. Sansone et al., 2008 unpublished manuscript). *See color plates.*

Notch and EGFr signaling for survival.[22,26,27,76–78] It was indeed proposed that gp130 and EGFr cross-talk involves an intracellular mediator,[73] and we observed that the intracellular fragment of Notch-3 was enough to induce the overexpression of Jagged-1 and CA-IX[26,27] (Figure 4–2). The cross-talk between Notch-3 and the EGFr pathway has been described in human cancer cells.[79] Furthermore, independent observation revealed that Notch acts upstream of ERK and Ras activation in (breast) cancer cell lines, leading us to suppose that Notch amplifies the activity of the entire EGFr/Ras pathway.[79–81] EGFr is of primary importance for MS survival, MS formation sustained by EGF supplementation to serum-free media, and the inhibition of both EGFr and Notch activation sufficient to halt MS formation.[22,26,27,77] Moreover, Notch-3 is of crucial importance for the proliferation and differentiation of human mammary gland progenitor cells and for the survival of breast cancer cells.[82,83] The overall picture suggests that IL-6 may act as an amplifier of the prosurvival activity of the Notch/EGFr/ERK axis.[84,85]

The IL-6/Notch-3 interplay was found to promote several malignant features in MS such as survival (via the expression of Jagged-1, a Notch ligand[26,27]), invasive capability, and hypoxia survival (via the upregulation of CA-IX, a transmembrane enzyme that catalyzes the hydration of carbon dioxide and balances intra/extracellular pH[86,87]). MS from aggressive ductal breast carcinoma show high IL-6 expression compared to MS obtained from the same individual's normal tissue.[26,27] It was therefore proposed that an autocrine IL-6–dependent loop promotes malignancy in breast cancer (stem) cells. Overall, these data provide a functional meaning to the basal/stem cell–like phenotype and support its functional association with the aggressive behavior of breast cancer cells and basal-like tumors.[42–45] Indirect confirmation of the inherent association occurring between the stem cell–like phenotype and the biologically aggressive behavior of breast cancer cells recently came from a study on the inflammatory breast carcinoma, an aggressive breast cancer subtype in which plenty of lymphovascular tumor emboli in the surrounding tumor tissue are present, and which has been postulated to arise from a basal-like breast cancer stem cell.[88–90] Indeed, cells capable of

Breast carcinoma lymphovascular tumor emboli

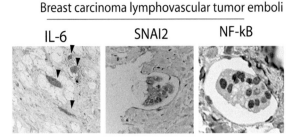

Figure 4–3: Immunohistochemical analysis of lymphovascular tumor emboli. *See color plates.*

generating such MS-like structures harbored in the lymphatic vessels, which represent a highly invasive breast cancer cell population, were found to over-express IL-6 (Figure 4–3), Notch-3,[90] CA-IX,[91] and a CD44+ stem cell–like pheno-type.[90] These data reinforce the notion that IL-6/Notch-3 expression associates with an invasive stem cell–like phenotype that is peculiarly represented in some breast cancer subtypes such as the basal-like and the inflammatory carcinoma of the breast.

SNAI2 GENE EXPRESSION IN BASAL/STEM CELL–LIKE CELLS: AN UNPREDICTABLE LINK WITH INFLAMMATION

The expression of SNAI gene family members triggers the process of epithelial to mesenchyme transition and the expression of a mesenchymal phenotype in epithelial cells.[57,92–94] It has been reported that mesenchymal breast tumors dis-close a basal-like phenotype[46] and that the expression of mesenchymal and SNAI genes is upregulated in basal-like tumor tissues and cell lines.[39,47,51,60,63,95] In vitro, the expression of SNAI family genes leads to the gain of a basal/stem cell–like phenotype and promotes stem cell features and MS-forming capacity in mammary gland stem cells.[28,50,51,58] These data reinforce the tight connec-tion between the basal/stem cell–like and mesenchymal phenotypes (Figure 4–1). We and other authors observed that vimentin and SNAI2 upregulation is present in MS[28,50] and that such an expression is reduced in MS-derived adherent cells.[24] Moreover, the transient SNAI2 knockdown was found to reduce ERα expression (a gene that is directly repressed by SNAI family genes[96]) and to upregulate CA-IX expression in breast cancer cells.[28] Furthermore, the stable SNAI2 knockdown also downregulates CD44, CK-5, and Jagged-1 expression (G. Storci, 2009, manuscript in preparation). These data agree with recent literature showing an overlap between the acquisition of a basal/stem and a mesenchy-mal phenotype in breast cancer cells.[51,58] The SNAI2 gene is a gene repressor, and it may therefore be proposed that the upregulation of the basal/stem cell–like phenotype may involve an indirect mechanism. Recent data indicate that the downregulation of E-Cadherin that allows β-catenin delocalization from the plasma membrane[24,97] and its nuclear activity represents a good candi-date to promote a mesenchymal phenotype and aggressiveness in breast cancer cells.[98–101] Indeed, CK5, CD44, Jagged-1, and CA-IX are targets of β-catenin, and

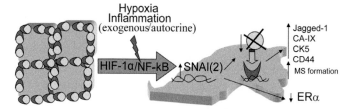

Figure 4–4: Schematic of the relationship between the hypoxia/inflammation-dependent upregulation of SNAI gene family members and the basal/stem cell–like phenotype: the repression of E-Cadherin unleashes β-catenin for promoter activation. Notably, components of the IL-6–driven pathway described in Figure 2 have been reported to induce Epithelial-mesenchyme transition (EMT) in various models.[58,63,105,106] *See color plates.*

SNAI2 expression[102–104] (G. Storci, 2009, manuscript in preparation) (Figure 4–4). Notably, β-catenin transgenic mice show a high prevalence of basal-like tumors, and human basal-like tumors show a peculiar cytoplamic localization of β-catenin, particularly in the mesenchymal cell component.[47,107] Moreover, it has been demonstrated that β-catenin, a widely acknowledged stem cell regulatory pathway, promotes mammary gland tumorigenesis by targeting stem/progenitor cells.[108,109] As observed for the IL-6-controlled network, SNAI2 expression promotes the invasive capability of MS, even those obtained from the normal mammary tissue, suggesting that SNAI expression may modulate the matrix-invasive capacity of mammary gland stem cells in physiologic conditions. In keeping with the hypothesis that the stem cell phenotype is inherently associated with aggressive features,[55,110,111] the expression of SNAI genes has been associated with breast cancer cells' invasive potential,[28,112,113] and the overexpression of SNAI2 was observed in lymphovascular tumor emboli (Figure 4–3). Similarly to what occurs to IL-6, we observed that SNAI2 upregulation in breast cancer cells is elicited by hypoxia exposure.[26,28] In breast cancer cells, IL-6 and SNAI2 were found also to be upregulated on exposure to pro-inflammatory molecules (such as tumor necrosis factor alpha) throughout a nuclear factor-kappaB (NF-κB)-dependent mechanism (G. Storci, 2009, manuscript in preparation).[114] Accordingly, the exposure to hypoxia and inflammatory stimuli elicits an epithelial to mesenchymal phenotype transition.[28,63,106,114–119] These data reinforce the notion that hypoxic/inflammatory pathways may concur in the promotion of a basal/stem cell–like phenotype in breast cancer cells (Figure 4–4).

THE INFLAMMATORY/HYPOXIC PROFILE IN MAMMARY GLAND STEM CELL REGULATION

NF-κB–regulated genes play a fundamental role in mammary gland morphogenesis, thus pointing out a primary role in stem cell regulation.[120,121] It was recently observed that the inhibition of NF-κB activity halts MS formation from mouse and human mammary gland[122] (G. Storci, 2009, manuscript in preparation). Similarly to what occurs to hemopoietic stem cells,[123] an overrepresentation of NF-κB–regulated genes in CD44$^+$ breast cancer stem cells was found.[29] Of note, the overexpression of the NF-κB activator Aurora kinase A and of NF-κB–regulated genes was found in basal-like and inflammatory tumors.[124–126] Moreover, an overexpression of nuclear NF-κB was observed in lymphovascular tumor emboli (Figure 4–3). The upregulation of NF-κB–regulated targets in CD44$^+$ breast cancer stem cells may be functionally linked with the overexpression of hypoxia-induced factor 1-alpha (HIF-1α) in such cells, in the absence of a hypoxic environment.[40] Indeed, a wealth of data indicates that NF-κB triggers the expression of the HIF-1α gene and that SNAI2 gene expression is under the control of both NF-κB and HIF-1α activation.[122,127–132] Interestingly, HIF-1α gene expression becomes upregulated also by Stat-3 activation.[133] Moreover, HIF-1α is upregulated in basal-like tumors, and HIF1-α and NF-κB are overexpressed in inflammatory breast cancer.[126,134,135] These data reinforce the hypothesis that an inflammatory/hypoxic response machinery is upregulated in breast cancer stem cells and in breast cancer subtypes that disclose overt stem cell–like features.

The upregulation of IL-6 and SNAI2 gene expression on hypoxia exposure in breast cancer cells and MS may be regarded as a feature of the normal regulation of stem cells. Indeed, literature data indicate that stem/progenitor cells of different origins are better expanded in vitro under hypoxic conditions and that hypoxia exposure upregulated stem cell features in differentiated cells.[27,136–140] Moreover, it has been proposed that the niches in which stem cells are harbored are localized in hypoxic tissue regions.[27,136,141] Notably, in the presence of hypoxia, Notch and β-catenin proteins physically interact with HIF-1α.[104,142] Hence the overactivation of hypoxia response pathways in stem cells may be under the control of inflammatory cues, leading us to propose that not only a hypoxic,[27,136,142,143] but also a pro-inflammatory environment may promote stem cell survival in the stem cell niche. This hypothesis may be considered together with the one suggesting that the mammary gland takes part in an ancient evolutionarily conserved inflammatory response machinery that has lately evolved into a feed-producing device.[144] Epidemiological and experimental data point out that inflammation and NF-κB pathway upregulation play a key role in mammary gland tumorigenesis.[145,146] The available data also indicate that pro-inflammatory factors, released from distant sites, as well as local inflammatory cells promote mammary gland tumorigenesis and aggressiveness.[147–149] Moreover, some models of chemically induced mammary tumors were recently revealed to be the consequence of a local inflammatory response.[150] These data indicate that the pro-inflammatory environment is of pivotal importance in mammary gland tumorigenesis. A peculiar situation may occur during aging, a major risk factor for breast cancer as well as a widely acknowledged pro-inflammatory condition.[151–155] Indeed, experimental models revealed that senescent stromal cells express increased levels of inflammatory molecules, such as IL-6, and promote tumor xenograft growth, a phenomenon that can be hampered by anti-IL6-specific antibody administration.[71,156,157] In this regard, we found that the expression of IL-6 and NF-κB is increased in aged

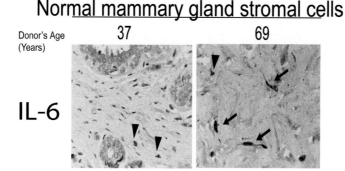

Figure 4–5: Immunohistochemical analysis of normal mammary gland stromal cells (arrowhead, macrophages; arrows, fibroblasts). *See color plates.*

mammary gland stromal cells (G. Storci, 2009, manuscript in preparation) (Figure 4–5). In keeping with this reasoning, it was proposed that mammary gland senile involution, a major risk factor for breast cancer, is molecularly analogous to an inflammatory process.[158] The overall emerging picture suggests that the upregulation of hypoxic/inflammatory pathways may confer survival on cancer stem cells and that an aged environment may facilitate the phenomenon (Figure 4–6).

CHEMORESISTANCE: A CONSEQUENCE OF THE UPREGULATION OF A HYPOXIC/INFLAMMATORY RESPONSE PATHWAY IN STEM CELLS

It may be reasoned that hypoxia is the most suitable place to store normal (mammary gland) stem cells because such an environment would keep normal stem cells away from oxidative damage, thus preserving DNA integrity throughout rounds of replications. Incidentally, the upregulation of the hypoxia response is likely to render normal stem cells resistant to cell death stimuli. Indeed, the presence of hypoxia invariably associates with radio- and chemoresistance in tumors and cancer cells.[143,160] Hence it is not unexpected that breast cancer stem cells that overexpress such genes are more chemoresistant than other tumor cells. Literature data indicate that a variety of genes expressed in stem cells are linked to enhanced resistance to cell death: IL-6,[161] SNAI2,[162] CD44,[163] Notch-3,[83] and NF-κB,[164] among others. Indeed, the stem cell–like phenotype is linked with resistance to cytotoxic agents and radiation.[52,56,163,165–169] Furthermore, the presence of CD44-positive and MS-generating cells in breast cancer tissues increases after chemotherapy.[169] It may therefore be proposed that hypoxia survival pathway activation increases chemoresistance in breast cancer cells. An acknowledged example of such a phenomenon is the expression of breast cancer resistance protein, an IL-6–regulated protein that extrudes mitochondrial protoporphirines (thus promoting hypoxia survival), chemotherapeutic drugs (chemoresistance), and the vital dye Hoechst 32352.[161,170,171] The exclusion of the dye leads to the side population phenotype that, in various tissues, including

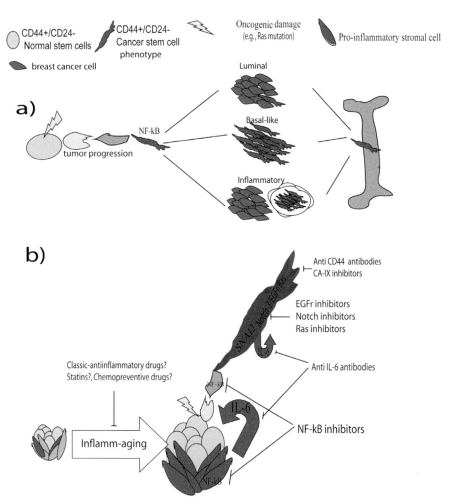

Figure 4–6: (a) Schematic of the basal/stem cell–like phenotype in various breast cancer subtypes. Disseminated breast cancer cells have been proposed to disclose a stem cell-like phenotype.[159] (b) The contribution of an aged/pro-inflammatory stromal cell environment (termed *inflamm-aging*[153]). A number of drugs, alone or in combination, may target cancer stem cells, and some of them may also be candidates for acting on the pro-tumorigenic environment.[71] *See color plates.*

the mammary gland and breast cancer cell lines, is enriched in cells showing a basal/stem cell–like phenotype, MS formation capability, and tumor-forming capacity.[21,165,172] The link between hypoxia survival and stem cell regulation also involves β-catenin, whose expression associates with radioresistance and increases in breast cancer stem cells exposed to cytotoxic drugs.[100,166] Hence there are hints suggesting that an inherent association between the stem cell phenotype and resistance to cancer therapy agents may be explained by the activation of pathways that are involved in stem cell maintenance. Of note, it has been proposed that NF-κB–targeting drugs are able to kill stem cells at a higher rate than their progeny.[123,173,174] Moreover, cancer stem cells have been found to be more sensitive than normal cells toward NF-κB inactivation.[123,175] Such data are in agreement with the proposed upregulation of inflammatory pathways in cancer stem cells, leading to speculation that the acknowledged anticancer properties of some anti-inflammatory drugs may be related to their preferential targeting of cancer stem cells. It may be proposed that molecules that can target specific components of such an hypoxic/inflammatory pathway may (alone or in combination) target stem cells and also pro-inflammatory/aged stromal cells[85,175,176] (Figure 4–6b). Moreover, this approach may be successfully applied to breast cancer subtypes in which the stem cell phenotype is particularly overt such as the basal-like and the inflammatory breast carcinoma[126,177] (Figure 4–6).

ACKOWLEDGMENTS

The author would like to thank Pasquale Sansone, Gianluca Storci, and Claudio Ceccarelli for their invaluable contribution to the past and current research.

REFERENCES

1. Hoshino K. Transplantability of mammary gland in brown fat pads of mice. *Nature.* 1967 Jan 14;213(5072):194–5.
2. Kordon EC and Smith GH. An entire functional mammary gland may comprise the progeny from a single cell. *Development.* 1998 May;125(10):1921–30.
3. Shackleton M, et al. Generation of a functional mammary gland from a single stem cell. *Nature.* 2006 Jan 5;439(7072):84–8.
4. Stingl J, et al. Purification and unique properties of mammary epithelial stem cells. *Nature.* 2006 Feb 23;439(7079):993–7.
5. Tsai YC, et al. Contiguous patches of normal human mammary epithelium derived from a single stem cell: implications for breast carcinogenesis. *Cancer Res.* 1996 Jan 15;56(2):402–4.
6. Diallo R, et al. Monoclonality in normal epithelium and in hyperplastic and neoplastic lesions of the breast. *J Pathol.* 2001 Jan;193(1):27–32.
7. Boecker W, et al. Evidence of progenitor cells of glandular and myoepithelial cell lineages in the human adult female breast epithelium: a new progenitor (adult stem) cell concept. *Cell Prolif.* 2003 Oct;36(Suppl 1):73–84.
8. Korsching E, et al. Cytogenetic alterations and cytokeratin expression patterns in breast cancer: integrating a new model of breast differentiation into cytogenetic pathways of breast carcinogenesis. *Lab Invest.* 2002 Nov;82(11):1525–33.
9. Villadsen R, et al. Evidence for a stem cell hierarchy in the adult human breast. *J Cell Biol.* 2007 Apr 9;177(1):87–101.
10. Gudjonsson T, et al. Isolation, immortalization, and characterization of a human breast epithelial cell line with stem cell properties. *Genes Dev.* 2002 Mar 15;16(6):693–706.
11. Stingl J and Caldas C. Molecular heterogeneity of breast carcinomas and the cancer stem cell hypothesis. *Nat Rev Cancer.* 2007 Oct;7(10):791–9.
12. Sleeman KE, et al. Dissociation of estrogen receptor expression and in vivo stem cell activity in the mammary gland. *J Cell Biol.* 2007 Jan 1;176(1):19–26.
13. Vaillant F, et al. The emerging picture of the mouse mammary stem. *Stem Cell Rev.* 2007 Jun;3(2):114–23.
14. Asselin-Labat ML, et al. Steroid hormone receptor status of mouse mammary stem cells. *J Natl Cancer Inst.* 2006 Jul 19;98(14):1011–4.
15. Regan J and Smalley M. Prospective isolation and functional analysis of stem and differentiated cells from the mouse mammary gland. *Stem Cell Rev.* 2007 Jun;3(2):124–36.

16. Nielsen TO, et al. Immunohistochemical and clinical characterization of the basal-like subtype of invasive breast carcinoma. *Clin Cancer Res*. 2004 Aug 15;10(16):5367–74.

17. Fulford LG, et al. Specific morphological features predictive for the basal phenotype in grade 3 invasive ductal carcinoma of breast. *Histopathology*. 2006 Jul;49(1):22–34.

18. Rakha EA, et al. Impact of basal-like breast carcinoma determination for a more specific therapy. *Pathobiology*. 2008 Jun;75(2):95–103.

19. Livasy CA, et al. Phenotypic evaluation of the basal-like subtype of invasive breast carcinoma. *Mod Pathol*. 2006 Feb;19(2):264–71.

20. Reis-Filho JS, et al. Triple negative tumours: a critical review *Histopathology*. 2008 Jan;52(1):108–18.

21. Dontu G, et al. In vitro propagation and transcriptional profiling of human mammary stem/progenitor cells. *Genes Dev*. 2003 May 15;17(10):1253–70.

22. Dontu G, et al. Role of Notch signaling in cell-fate determination of human mammary stem/progenitor cells. *Breast Cancer Res*. 2004 Aug;6(6):R605–15.

23. Dontu G, and Wicha MS. Survival of mammary stem cells in suspension culture: implications for stem cell biology and neoplasia. *J Mammary Gland Biol Neoplasia*. 2005 Jan;10(1):75–86.

24. Dontu G, et al. Stem cells in mammary development and carcinogenesis: implications for prevention and treatment. *Stem Cell Rev*. 2005;1(3):207–13.

25. Liu S, et al. BRCA1 regulates human mammary stem/progenitor cell fate. *Proc Natl Acad Sci U S A*. 2008 Feb 5;105(5):1680–5.

26. Sansone P, et al. IL-6 triggers malignant features in mammospheres from human ductal breast carcinoma and normal mammary gland. *J Clin Invest*. 2007 Dec;117(12):3988–4002.

27. Sansone P, et al. p66Shc/Notch-3 interplay controls self-renewal and hypoxia survival in human stem/progenitor cells of the mammary gland expanded in vitro as mammospheres. *Stem Cells*. 2007 Mar;25(3):807–15.

28. Storci G, et al. The basal-like breast carcinoma phenotype is regulated by SLUG gene expression. *J Pathol*. 2008 Jan;214(1):25–37.

29. Liu R, et al. The prognostic role of a gene signature from tumorigenic breast-cancer cells. *N Engl J Med*. 2007 Jan 18;356(3):217–26.

30. Bertucci F, et al. A gene signature in breast cancer. *N Engl J Med*. 2007 May 3;356(18):1887–8.

31. Ben-Porath I, et al. An embryonic stem cell-like gene expression signature in poorly differentiated aggressive human tumors. *Nat Genet*. 2008 May;40(5):499–507.

32. Nakshatri H, et al. Breast cancer stem cells and intrinsic subtypes: controversies rage on. *Curr Stem Cell Res Ther*. 2009 Jan;4(1):50–60.

33. James CR, et al. BRCA1, a potential predictive biomarker in the treatment of breast cancer. *Oncologist*. 2007 Feb;(12):142–50.

34. Herschkowitz JI, et al. Identification of conserved gene expression features between murine mammari carcinoma models and human breast tumors. *Genome Biol*. 2007 May;8(5):R76.

35. Al-Hajj M, et al. Prospective identification of tumorigenic breast cancer cells. *Proc Natl Acad Sci U S A*. 2003 Apr 1;100(7):3983–88.

36. Cobaleda C, et al. The emerging picture of human breast cancer as a stem cell-based disease. *Stem Cell Rev*. 2008 Summer;4(2):67–79.

37. Wicha MS, et al. Cancer stem cells: an old idea – a paradigm shift. *Cancer Res*. 2006 Feb 15;66(4):1883–90.

38. Honeth G, et al. The CD44+/CD24- phenotype is enriched in basal-like breast tumors. *Breast Cancer Res*. 2008 Jun;10(3):R53.

39. Charafe-Jauffret E, et al. Gene expression profiling of breast cell lines identifies potential new basal markers. *Oncogene*. 2006 Apr 6;25(15):2273–84.

40. Shipitsin M, et al. Molecular definition of breast tumor heterogeneity *Cancer Cell*. 2007 Mar;11(3):259–73.

41. Bertucci F, et al. How different are luminal A and basal breast cancers? *Int J Cancer.* 2009 Mar 15;124(6):1338–48.

42. Reedijk M, et al. High-level coexpression of JAG1 and NOTCH1 is observed in human breast cancer and is associated with poor overall survival. *Cancer Res.* 2005 Sep 15; 65(18):8530–7.

43. Reedijk M, et al. JAG1 expression is associated with a basal phenotype and recurrence in lymph node-negative breast cancer. *Breast Cancer Res Treat.* 2008 Oct;111(3):439–48.

44. Dickson BC, et al. High-level JAG1 mRNA and protein predict poor outcome in breast cancer. *Mod Pathol.* 2007 Jun;20(6):685–93.

45. Brennan DJ, et al. CA IX is an independent prognostic marker in premenopausal breast cancer patients with one to three positive lymph nodes and a putative marker of radiation resistance. *Clin Cancer Res.* 2006 Nov;12(21):439–48.

46. Kuroda N, et al. Basal-like carcinoma of the breast: further evidence of the possibility that most metaplastic carcinomas may be actually basal-like carcinomas. *Med Mol Morphol.* 2008 Jun;41(2):117–20.

47. Sarrió D, et al. Epithelial-mesenchymal transition in breast cancer relates to the basal-like phenotype. *Cancer Res.* 2008 Feb 15;68(4):989–97.

48. Rakha EA, et al. Breast carcinoma with basal differentiation: a proposal for pathology definition based on basal cytokeratin expression. *Histopathology.* 2007 Mar;50(4): 434–8.

49. Sleeman KE, et al. Dissociation of estrogen receptor expression and in vivo stem cell activity in the mammary gland. *J Cell Biol.* 2007 Jan 1;176(1):19–26.

50. Liao MJ, et al. Hypoxia-inducible factor-1alpha is a key regulator of metastasis in a transgenic model of cancer initiation and progression. *Cancer Res.* 2007 Sep 1;67(17):8131–8.

51. Mani SA, et al. The epithelial-mesenchymal transition generates cells with properties of stem Cells. *Cell.* 2008 May 16;133(4):704–15.

52. Phillips TM, et al. The response of CD24(–/low)/CD44+ breast cancer-initiating cells to radiation. *J Natl Cancer Inst.* 2006 Dec 20;98(24):1777–85.

53. Ponti D, et al. Isolation and in vitro propagation of tumorigenic breast cancer cells with stem/progenitor cell properties. *Cancer Res.* 2005 Jul 1;65(13):5506–11.

54. Horwitz KB, et al. Rare steroid receptor-negative basal-like tumorigenic cells in luminal subtype human breast cancer xenografts. *Proc Natl Acad Sci U S A.* 2008 Apr 15;105(15):5774–9.

55. Sheridan C, et al. CD44+/CD24– breast cancer cells exhibit enhanced invasive properties: an early step necessary for metastasis. *Breast Cancer Res.* 2006;8(5):R59.

56. Fillmore CM, et al. Human breast cancer cell lines contain stem-like cells that self-renew, give rise to phenotypically diverse progeny and survive chemotherapy. *Breast Cancer Res.* 2008 Mar;10(2):R25.

57. Bolós V, et al. The transcription factor Slug represses E-cadherin expression and induces epithelial to mesenchymal transitions: a comparison with Snail and E47 repressors. *J Cell Sci.* 2003 Feb 1;116(Pt 3):499–511.

58. Morel AP, et al. Generation of breast cancer stem cells through epithelial-mesenchymal transition. *PLoS ONE.* 2008 Aug 6;3(8):e2888.

59. Gordon LA, et al. Breast cell invasive potential relates to the myoepithelial phenotype *Int J Cancer.* 2003 Aug 10;106(1):8–16.

60. Neve RM, et al. A collection of breast cancer cell lines for the study of functionally distinct cancer subtypes. *Cancer Cell.* 2006 Dec;10(6):515–27.

61. Hugo H, et al. Epithelial – mesenchymal and mesenchymal – epithelial transitions in carcinoma progression. *J Cell Physiol.* 2007 Nov;213(2):374–83.

62. Korsching E, et al. The origin of vimentin expression in invasive breast cancer: epithelial-mesenchymal transition, myoepithelial histogenesis or histogenesis from progenitor cells with bilinear differentiation potential. *J Pathol.* 2005 Aug;206(4):451–7.

63. Blick T, et al. Epithelial mesenchymal transition traits in human breast cancer cell lines. *Clin Exp Metastasis.* 2008 May;25(6):629–42.

64. Knüpfer H and Preiss R. Significance of interleukin-6 (IL-6) in breast cancer. *Breast Cancer Res Treat*. 2007 Apr;102(2):129–35.

65. Lien HC, et al. Molecular signatures of metaplastic carcinoma of the breast by large-scale transcriptional profiling: identification of genes potentially related to epithelial–mesenchymal transition. *Oncogene*. 2007 Dec 13;26(57):7859–71.

66. Sasser AK, et al. Interleukin-6 is a potent growth factor for ER-alpha-positive human breast cancer. *FASEB J*. 2007 Nov;21(13):3763–70.

67. Ancrile B, et al. Oncogenic Ras-induced secretion of IL6 is required for tumorigenesis. *Genes Dev*. 2007 Jul 15;21(14):1714–19.

68. Bolós V, et al. Notch signaling in development and cancer. *Endocr Rev*. 2007 May;28(3):339–63.

69. Tan EY, et al. The key hypoxia regulated gene CAIX is upregulated in basal-like breast tumours and is associated with resistance to chemotherapy. *Br J Cancer*. 2009 Jan;100(2):405–11.

70. Gao SP, et al. Mutations in the EGFR kinase domain mediate STAT3 activation via IL-6 production in human lung adenocarcinomas. *J Clin Invest*. 2007 Dec;117(12):3846–56.

71. Studebaker AW, et al. Fibroblasts isolated from common sites of breast cancer metastasis enhance cancer cell growth rates and invasiveness in an interleukin-6-dependent manner. *Cancer Res*. 2008 Nov;68(21):9087–95.

72. Hoadley KA, et al. EGFR associated expression profiles vary with breast tumor subtype. *BMC Genomics*. 2007 Jul 31;8:258.

73. Zhao L, et al. Mammary gland remodeling depends on gp130 signaling through Stat3 and MAPK. *J Biol Chem*. 2004 Oct 15;279(42):44093–100.

74. Viti J, et al. Epidermal growth factor receptors control competence to interpret leukemia inhibitory factor as an astrocyte inducer in developing cortex. *J Neurosci*. 2003 Apr 15;23(8):3385–93.

75. Shimazaki T, et al. The ciliary neurotrophic factor/leukemia inhibitory factor/gp130 receptor complex operates in the maintenance of mammalian forebrain neural stem cells. *J Neurosci*. 2001 Oct 1;21(19):7642–53.

76. Reynolds BA and Rietze RL. Neural stem cells and neurospheres – re-evaluating the relationship. *Nat Methods*. 2005 May;2(5):333–6.

77. Farnie GJ, et al. Novel cell culture technique for primary ductal carcinoma in situ: role of Notch and epidermal growth factor receptor signaling pathways. *J Natl Cancer Inst*. 2007 Apr 18;99(8):616–27.

78. Chojnacki A, et al. Glycoprotein 130 signaling regulates Notch1 expression and activation in the self-renewal of mammalian forebrain neural stem cells. *J Neurosci*. 2003 Mar 1;23(5):1730–41.

79. Haruki N, et al. Dominant-negative Notch3 receptor inhibits mitogen-activated protein kinase pathway and the growth of human lung cancers. *Cancer Res*. 2005 May 1;65(9):3555–61.

80. Talora C, et al. Cross talk among Notch3, pre-TCR, and Tal1 in T-cell development and leukemogenesis. *Blood*. 2006 Apr 15;107(8):3313–20.

81. Fitzgerald K, et al. Ras pathway signals are required for notch-mediated oncogenesis. *Oncogene*. 2000 Aug 31;19(37):4191–8.

82. Raouf A, et al. Transcriptome analysis of the normal human mammary cell commitment and differentiation process. *Cell Stem Cell*. 2008 Jul 3;3(1):109–18.

83. Yamaguchi N, et al. NOTCH3 signaling pathway plays crucial roles in the proliferation of ErbB2-negative human breast cancer cells. *Cancer Res*. 2008 Mar 15;68(6):1881–8.

84. Grivennikov S and Karin M. Autocrine IL-6 signaling: a key event in tumorigenesis? *Cancer Cell*. 2008 Jan;13(1):7–9.

85. Schafer ZT, and Brugge JS, IL-6 involvement in epithelial cancers. *J Clin Invest*. 2007 Dec;117(12):3660–3.

86. Hilvo M, et al. Biochemical characterization of CA IX, one of the most active carbonic anhydrase isozymes. *J Biol Chem*. 2008 Oct, 283(41): 27799–809.

87. Supuran CT. Carbonic anhydrases: novel therapeutic applications for inhibitors and activators. *Nat Rev Drug Discov*. 2008 Feb;7(2):168–81.

88. Dirix LY, et al. Inflammatory breast cancer: current understanding. *Curr Opin Oncol*. 2006 Nov;18(6):563–71.

89. Ben Hamida A, et al. Markers of subtypes in inflammatory breast cancer studied by immunohistochemistry: prominent expression of P-cadherin. *BMC Cancer*. 2008 Jan 29;8:28.

90. Xiao Y, et al. The lymphovascular embolus of inflammatory breast cancer expresses a stem cell-like phenotype. *Am J Pathol*. 2008 Aug;173(2):561–74.

91. Colpaert CG, et al. Inflammatory breast cancer shows angiogenesis with high endothelial proliferation rate and strong E-cadherin expression. *Br J Cancer*. 2003 Mar 10;88(5):718–25.

92. Moreno-Bueno G, et al. Genetic profiling of epithelial cells expressing E-cadherin repressors reveals a distinct role for Snail, Slug, and E47 factors in epithelial–mesenchymal transition. *Cancer Res*. 2006 Oct 1;66(19):9543–56.

93. Hajra KM, et al. The SLUG zinc-finger protein represses E-cadherin in breast cancer. *Cancer Res*. 2002 Mar 15;62(6):1613–8.

94. Thiery JP and Sleeman JP. Complex networks orchestrate epithelial-mesenchymal transitions. *Nat Rev Mol Cell Biol*. 2006 Feb;7(2):131–42.

95. Zajchowski DA, et al. Identification of gene expression profiles that predict the aggressive behaviour of breast cancer cells. *Cancer Res*. 2001 Jul 1;61(13):5168–78.

96. Dhasarathy A, et al. The transcription factor snail mediates epithelial to mesenchymal transitions by repression of estrogen receptor-alpha. *Mol Endocrinol*. 2007 Dec;21(12):2907–18.

97. Nelson WJ, et al. Convergence of Wnt, beta-catenin, and cadherin pathways. *Science*. 2004 Mar 5;303(5663):1483–7.

98. Leong KG, et al. Jagged1-mediated Notch activation induces epithelial-to-mesenchymal transition through Slug-induced repression of E-cadherin. *J Exp Med*. 2007 Nov 26;204(12):2935–48.

99. Prasad CP, et al. Epigenetic alterations of CDH1 and APC genes: relationship with activation of Wnt/beta-catenin pathway in invasive ductal carcinoma of breast. *Life Sci*. 2008 Aug;83(9–10):318–25.

100. Woodward WA, et al. WNT/beta-catenin mediates radiation resistance of mouse mammary progenitor cells. *Proc Natl Acad Sci U S A*. 2007 Jan 9;104(2):618–23.

101. Lindvall C, et al. Wnt signaling, stem cells, and the cellular origin of breast cancer. *Stem Cell Rev*. 2007 Jun;3(2):157–68.

102. Schwartz DR, et al. Novel candidate targets of beta-catenin/T-cell factor signaling identified by gene expression profiling of ovarian endometrioid adenocarcinomas. *Cancer Res*. 2003 Jun 1;63(11):2913–22.

103. Estrach S, et al. Jagged 1 is a beta-catenin target gene required for ectopic hair follicle formation in adult epidermis. *Development*. 2006 Nov;133(22):4427–38.

104. Kaidi A, et al. Interaction between beta-catenin and HIF-1 promotes cellular adaptation to hypoxia. *Nat Cell Biol*. 2007 Feb;9(2):210–17.

105. Lo HW, et al. Epidermal growth factor receptor cooperates with signal transducer and activator of transcription 3 to induce epithelial-mesenchymal transition in cancer cells via up-regulation of TWIST gene expression. *Cancer Res*. 2007 Oct 1;67(19):9066–76.

106. Sahlgren C, et al. Notch signaling mediates hypoxia-induced tumor cell migration and invasion. *Proc Natl Acad Sci U S A*. 2008 Apr 29;105(17):6392–7.

107. McCarthy A, et al. A mouse model of basal-like breast carcinoma with metaplastic elements. *J Pathol*. 2007 Mar;211(4):389–98.

108. Liu BY, et al. The transforming activity of Wnt effectors correlates with their ability to induce the accumulation of mammary progenitor cells. *Proc Natl Acad Sci U S A*. 2004 Mar 23;101(12):4158–63.

109. Li Y, et al. Evidence that transgenes encoding components of the Wnt signaling pathway preferentially induce mammary cancers from progenitor cells. *Proc Natl Acad Sci U S A*. 2003 Dec 23;100(26):15853–8.

110. Brabletz T, et al. Opinion: migrating cancer stem cells – an integrated concept of malignant tumour progression. *Nat Rev Cancer.* 2005 Sep;5(9):744–9.

111. Grimshaw MJ, et al. Mammosphere culture of metastatic breast cancer cells enriches for tumorigenic breast cancer cells. *Breast Cancer Res.* 2008;10(3):R52.

112. Olmeda D, et al. SNAI1 is required for tumor growth and lymph node metastasis of human breast carcinoma MDA-MB-231 cells. *Cancer Res.* 2007 Dec 15;67(24):11721–31.

113. Kurrey NK and Bapat SA. Snail and Slug are major determinants of ovarian cancer invasiveness at the transcription level. *Gynecol Oncol.* 2005;97:155–65.

114. Dong R, et al. Role of nuclear factor kappa B and reactive oxygen species in the tumor necrosis factor-alpha-induced epithelial-mesenchymal transition of MCF-7 cells. *Braz J Med Biol Res.* 2007 Aug;40(8):1071–8.

115. Huber MA, et al. NF-kappaB is essential for epithelial-mesenchymal transition and metastasis in a model of breast cancer progression. *J Clin Invest.* 2004 Aug;114(4):569–81.

116. Grund EM, et al. Tumor necrosis factor-alpha regulates inflammatory and mesenchymal responses via mitogen-activated protein kinase kinase, p38, and nuclear factor kappaB in human endometriotic epithelial cells. *Mol Pharmacol.* 2008 May;73(5):1394–404.

117. Lester RD, et al. uPAR induces epithelial-mesenchymal transition in hypoxic breast cancer cells. *J Cell Biol.* 2007 Jul 30;178(3):425–36.

118. Julien S, et al. Activation of NF-kappaB by Akt upregulates Snail expression and induces epithelium mesenchyme transition. *Oncogene.* 2007 Nov 22;26(53):7445–56.

119. Chua HL, et al. NF-kappaB represses E-cadherin expression and enhances epithelial to mesenchymal transition of mammary epithelial cells: potential involvement of ZEB-1 and ZEB-2. *Oncogene.* 2007 Feb 1;26(5):711–24.

120. Brantley DM, et al. Nuclear factor-kappaB (NF-kappaB) regulates proliferation and branching in mouse mammary epithelium. *Mol Biol Cell.* 2001 May;12(5):1445–55.

121. Demicco EG, et al. RelB/p52 NF-kappaB complexes rescue an early delay in mammary gland development in transgenic mice with targeted superrepressor IkappaB-alpha expression and promote carcinogenesis of the mammary gland. *Mol Cell Biol.* 2005 Nov;25(22):10136–47.

122. Cao Y, et al. IkappaB kinase alpha kinase activity is required for self-renewal of ErbB2/Her2-transformed mammary tumor-initiating cells. *Proc Natl Acad Sci U S A.* 2007 Oct 2;104(40):15852–7.

123. Guzman ML, et al. The sesquiterpene lactone parthenolide induces apoptosis of human acute myelogenous leukemia stem and progenitor cells. *Blood.* 2005 Jun 1; 105(11):4163–9.

124. Finetti P, et al. Sixteen-kinase gene expression identifies luminal breast cancers with poor prognosis. *Cancer Res.* 2008 Feb 1;68(3):767–76.

125. Sun C, et al. Aurora kinase inhibition downregulates NF-kappaB and sensitises tumour cells to chemotherapeutic agents. *Biochem Biophys Res Commun.* 2007 Jan 5;352(1):220–5.

126. Van Laere SJ, et al. NF-kappaB activation in inflammatory breast cancer is associated with oestrogen receptor downregulation, secondary to EGFR and/or ErbB2 overexpression and MAPK hyperactivation. *Br J Cancer.* 2007 Sep 3;97(5):659–69.

127. Laffin B, et al. Loss of singleminded-2s in the mouse mammary gland induces an epithelial-mesenchymal transition associated with up-regulation of slug and matrix metalloprotease 2. *Mol Cell Biol.* 2008 Mar;28(6):1936–46.

128. Belguise K, et al. Green tea polyphenols reverse cooperation between c-Rel and CK2 that induces the aryl hydrocarbon receptor, slug, and an invasive phenotype. *Cancer Res.* 2007 Dec 15;67(24):11742–50.

129. Ikuta T, and Kawajiri K, Zinc finger transcription factor Slug is a novel target gene of aryl hydrocarbon receptor. *Exp Cell Res.* 2006 Nov 1;312(18):3585–94.

130. Belaiba RS, et al. Hypoxia up-regulates hypoxia-inducible factor-1alpha transcription by involving phosphatidylinositol 3-kinase and nuclear factor kappaB in pulmonary artery smooth muscle cells. *Mol Biol Cell.* 2007 Dec;18(12):4691–7.

131. Rius J, et al. NF-kappaB links innate immunity to the hypoxic response through transcriptional regulation of HIF-1alpha. *Nature*. 2008 Jun 5;453(7196):807–11.

132. Van Uden P, et al. Regulation of hypoxia-inducible factor-1alpha by NF-kappaB. *Biochem J*. 2008 Jun 15;412(3):477–84.

133. Xu Q, et al. Targeting Stat3 blocks both HIF-1 and VEGF expression induced by multiple oncogenic growth signaling pathways. *Oncogene*. 2005 Aug 25;24(36):5552–60.

134. Bertucci F, et al. Gene expression profiling for molecular characterization of inflammatory breast cancer and prediction of response to chemotherapy. *Cancer Res*. 2004 Dec;64(23):8558–65.

135. Van Den Eynden GG, et al. Gene expression profiles associated with the presence of a fibrotic focus and the growth pattern in lymph node-negative breast cancer. *Clin Cancer Res*. 2008 May 15;14(10):2944–52.

136. Simon MC and Keith B. The role of oxygen availability in embryonic development and stem cell function. *Nat Rev Mol Cell Biol*. 2008 Apr;9(4):285–96.

137. Cipolleschi MG, et al. The role of hypoxia in the maintenance of hematopoietic stem cells. *Blood*. 1993 Oct 1;82(7):2031–7.

138. Holmquist L, et al. Effect of hypoxia on the tumor phenotype: the neuroblastoma and breast cancer models. *Adv Exp Med Biol*. 2006;587:179–93.

139. Axelson H, et al. Hypoxia-induced dedifferentiation of tumor cells – a mechanism behind heterogeneity and aggressiveness of solid tumors. *Semin Cell Dev Biol*. 2005 Aug–Oct;16(4–5):554–63.

140. Jögi A, et al. Hypoxia alters gene expression in human neuroblastoma cells toward an immature and neural crest-like phenotype. *Proc Natl Acad Sci U S A*. 2002 May 14;99(10):7021–6.

141. Parmar K, et al. Distribution of hematopoietic stem cells in the bone marrow according to regional hypoxia. *Proc Natl Acad Sci U S A*. 2007 Mar 27;104(13):5431–6.

142. Gustafsson MV, et al. Hypoxia requires notch signaling to maintain the undifferentiated cell state. *Dev Cell*. 2005 Nov;9(5):617–28.

143. Das B, et al. Hypoxia enhances tumor stemness by increasing the invasive and tumorigenic side population fraction. *Stem Cells*. 2008 Jul;26(7):1818–30.

144. Vorbach C, et al. Evolution of the mammary gland from the innate immune system? *Bioessays*. 2006 Jun;28(6):606–16.

145. Romieu-Mourez R, et al. Mouse mammary tumor virus c-rel transgenic mice develop mammary tumors. *Mol Cell Biol*. 2003 Aug;23(16):5738–54.

146. Mantovani A, et al. Inflammation and cancer: breast cancer as a prototype. *Breast*. 2007 Dec;16(Suppl 2):27–33.

147. Rao VP, et al. Breast cancer: should gastrointestinal bacteria be on our radar screen? *Cancer Res*. 2007 Feb 1;67(3):847–50.

148. Rao VP, et al. Innate immune inflammatory response against enteric bacteria Helicobacter hepaticus induces mammary adenocarcinoma in mice. *Cancer Res*. 2006 Aug 1;66(15):7395–400.

149. Mantovani A, et al. Cancer-related inflammation. *Nature*. 2008 Jul 24;454(7203):436–44.

150. Shin SR, et al. 7,12-dimethylbenz(a)anthracene treatment of a c-rel mouse mammary tumor cell line induces epithelial to mesenchymal transition via activation of nuclear factor-kappaB. *Cancer Res*. 2006 Mar 1;66(5):2570–5.

151. Benz CC. Impact of aging on the biology of breast cancer. *Crit Rev Oncol Hematol*. 2008 Apr;66(1):65–74.

152. Balducci L and Ershler WB. Cancer and ageing: a nexus at several levels. *Nat Rev Cancer*. 2005 Aug;5(8):655–62.

153. Franceschi C, et al. Inflamm-aging. An evolutionary perspective on immunosenescence. *Ann N Y Acad Sci*. 2000 Jun;908:244–54.

154. Bonafè M, et al. A gender-dependent genetic predisposition to produce high levels of IL-6 is detrimental for longevity. *Eur J Immunol*. 2001 Aug;31(8):2357–61.

155. Ferrucci L, et al. The origins of age-related proinflammatory state. *Blood*. 2005 Mar 15;105(6):2294–9.
156. Krtolica A, et al. Senescent fibroblasts promote epithelial cell growth and tumorigenesis: a link between cancer and aging. *Proc Natl Acad Sci U S A*. 2001 Oct 9;98(21):12072–7.
157. Yang G, et al. The chemokine growth-regulated oncogene 1 (Gro-1) links RAS signaling to the senescence of stromal fibroblasts and ovarian tumorigenesis. *Proc Natl Acad Sci U S A*. 2006 Oct 31;103(44):16472–7.
158. Schedin P, et al. Microenvironment of the involuting mammary gland mediates mammary cancer progression. *J Mammary Gland Biol Neoplasia*. 2007 Mar;12(1):71–82.
159. Balic M, et al. Most early disseminated cancer cells detected in bone marrow of breast cancer patients have a putative breast cancer stem cell phenotype. *Clin Cancer Res*. 2006 Oct 1;12(19):5615–21.
160. Graeber TG, et al. Hypoxia-mediated selection of cells with diminished apoptotic potential in solid tumours. *Nature*. 1996;379:88–91.
161. Conze D, et al. Autocrine production of interleukin 6 causes multidrug resistance in breast cancer cells. *Cancer Res*. 2001 Dec 15;61(24):8851–8.
162. Vannini I, et al. Short interfering RNA directed against the SLUG gene increases cell death induction in human melanoma cell lines exposed to cisplatin and fotemustine. *Cell Oncol*. 2007;29(4):279–87.
163. Toole BP, et al. Hyaluronan: a constitutive regulator of chemoresistance and malignancy in cancer cells. *Semin Cancer Biol*. 2008 Aug;18(4):244–50.
164. Biswas DK, et al. NF-kappa B activation in human breast cancer specimens and its role in cell proliferation and apoptosis. *Proc Natl Acad Sci U S A*. 2004 Jul 6;101(27):10137–42.
165. Cariati M, et al. Alpha-6 integrin is necessary for the tumourigenicity of a stem cell-like subpopulation within the MCF7 breast cancer cell line. *Int J Cancer*. 2008 Jan 15; 122(2):298–304.
166. Chen MS, et al. Wnt/beta-catenin mediates radiation resistance of Sca1+ progenitors in an immortalized mammary gland cell line. *J Cell Sci*. 2007 Feb 1;120(Pt 3):468–77.
167. Shafee N, et al. Cancer stem cells contribute to cisplatin resistance in Brca1/p53-mediated mouse mammary tumors. *Cancer Res*. 2008 May 1;68(9):3243–50.
168. Li X, et al. Intrinsic resistance of tumorigenic breast cancer cells to chemotherapy. *J Natl Cancer Inst*. 2008 May 7;100(9):672–9.
169. Li HZ, et al. Suspension culture combined with chemotherapeutic agents for sorting of breast cancer stem cells. *BMC Cancer*. 2008 May 14;8:135.
170. Krishnamurthy P, et al. The stem cell marker Bcrp/ABCG2 enhances hypoxic cell survival through interactions with heme. *J Biol Chem*. 2004 Jun 4;279(23):24218–25.
171. Patrawala L, et al. Side population is enriched in tumorigenic, stem-like cancer cells, whereas ABCG2+ and ABCG2− cancer cells are similarly tumorigenic. *Cancer Res*. 2005 Jul 15;65(14):6207–19.
172. Eyler CE and Rich JN. Survival of the fittest: cancer stem cells in therapeutic resistance and angiogenesis. *J Clin Oncol*. 2008 Jun 10;26(17):2839–45.
173. Zhou J, et al. Cancer stem/progenitor cell active compound 8-quinolinol in combination with paclitaxel achieves an improved cure of breast cancer in the mouse model. *Breast Cancer Res Treat*. 2008 May 28.
174. Zhou J, et al. NF-kappaB pathway inhibitors preferentially inhibit breast cancer stem-like cells. *Breast Cancer Res Treat*. 2008 Oct;111(3):419–27.
175. Jordan CT, et al. Cancer stem cells. *N Engl J Med*. 2006 Sep 21;355(12):1253–61.
176. Schlotter CM, et al. Molecular targeted therapies for breast cancer treatment. *Breast Cancer Res*. 2008 Jul 24;10(4):211.
177. Singh S, et al. Nuclear factor-kappaB activation: a molecular therapeutic target for estrogen receptor-negative and epidermal growth factor receptor family receptor-positive human breast cancer *Mol Cancer Ther*. 2007 Jul;6(7):1973–82.

SECTION II: THERAPEUTIC IMPLICATIONS OF CANCER STEM CELLS

5 Preventative and therapeutic strategies for cancer stem cells

Stewart Sell
Wadsworth Center, Ordway Research Institute, and the University at Albany

Gennadi Glinsky
Ordway Research Institute

Treatment of cancer should be directed to cancer stem cells as well as to the stage of maturation arrest at which the cancer cells accumulate. Cancers contain the same cell populations as do normal adult tissues: stem cells, proliferating transit-amplifying cells, terminally differentiated (mature cells), and dead cells. During normal tissue renewal, the number of proliferating transit-amplifying cells is essentially the same as the number of terminally differentiating cells so that the total number of cells remains relatively constant. On the other hand, in cancer tissue, the transit-amplifying cells are arrested at a stage of maturation in which they continue to proliferate and accumulate so that the mass of cancerous tissue continues to increase. The ability of retinoic acids to induce differentiation of teratocarcinoma stem cells provided a proof of principle that cancer stem cells could be induced to differentiate (differentiation therapy). Differentiation therapy has been applied with great success to cancer of the blood cells (leukemias) by inactivation of the signaling pathways that allow the leukemic transit-amplifying cells to continue to proliferate and not die (maturation arrest). Conventional therapies, such as chemotherapy, radiotherapy, and antiangiogenic therapies, also act on the proliferating cancer transit-amplifying cells. When these therapies are discontinued, the cancer will reform from the therapy-resistant cancer stem cells. Cancer stem cell–directed therapy may be possible using small inhibitory molecules or inhibitory RNAs (iRNA) to block the signals that maintain stemness so that the cancer stem cells are allowed to differentiate. Successful differentiation therapy of cancer stem cells would force these cells to differentiate so that they could no longer reestablish the cancer. In conclusion, to be curative, cancer therapy has to act not only on the cancer transit-amplifying cells responsible for the growth of the cancer, but also on the cancer stem cells that can reconstitute the cancer after successful therapy of the cancer transit-amplifying cells.

CELL LINEAGES AND CANCER

Figure 5–1 shows a simplified diagram of a cell lineage. This could be the lineage of either normal tissue or of cancer. The tissue or cancer stem cell is the cell from which the other cells in the lineage are derived. Stem cells are retained in the tissue because when a stem cell divides, it gives rise to one daughter cell that remains a stem cell and one daughter cell that begins the process of determination. This is known as asymmetric division. The cells in normal tissue are continually replaced by the proliferating daughter cells (normal tissue renewal). During normal tissue renewal, the daughter cell that differentiates after division of the stem cell retains the capacity to divide for several divisions before the progeny of this daughter cell can no longer divide (terminal differentiation). These proliferating cells are known as transit-amplifying cells. Transit-amplifying cells maintain the population of cells in both normal and cancer tissue. The difference between normal tissue renewal and cancer growth is a balance between production of new cells and differentiating cells in normal tissue so that the normal tissue mass remains essentially constant, whereas in cancer tissue, there is a block in the maturation of the differentiation of cancer transit-amplifying cells. This results in the cells not differentiating as do normal tissues, but rather, they become arrested at a stage of development in which they continue to proliferate so that

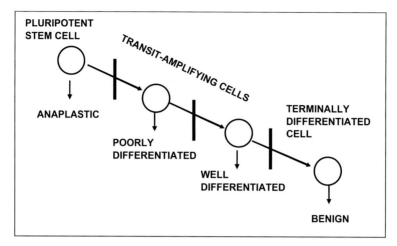

Figure 5–1: Cell lineage of cancer and normal tissue. Postulated stages of maturation arrest of a cell lineage and resultant cancer types. Depicted is a cell lineage from a totipotent stem cell to a terminally differentiated cell. In adult tissues, the most primitive stem cells normally do not divide. Normal tissue renewal is accomplished by proliferation of the transit-amplifying (TA) cells. TA cells in normal tissues have a finite lifetime so that the progeny of the TA become terminally differentiated cells and die by apoptosis. Cancers arise by blocks in the normal maturation process (maturation arrest). When this occurs, the TA cells continue to proliferate at the stage of maturation arrest. If this stage is early in the differentiation process, poorly differentiated cancers will be produced; if later, well-differentiated cancer or benign tumors are produced. Modified from Sell.[6]

the mass of the cancer tissue increases (maturation arrest). One of the ways to treat cancer is to force the malignant cancer-proliferating cells to differentiate into benign terminally differentiated cells.

Malignant cells can become benign

Malignant cells can differentiate into mature terminally differentiated cells. This was demonstrated by the classic pulse-labeling studies of Pierce and Wallace in 1971.[1] They showed that, during growth of a squamous cell carcinoma, the proliferating cancer cells became terminally differentiated keratin-containing cells. They pulse-labeled the newly synthesized DNA of proliferating squamous carcinoma cells with tritiated thymidine. Immediately after labeling, only the dividing cancer cells were labeled. Several days later, the terminally differentiated cells of the cancer, and even the keratin produced by the cancer, contained the label. If malignant cells can become benign, it should be possible to treat not only the cancer transit-amplifying cells, but also the cancer stem cells by forcing them to differentiate.

Differentiation therapy

The basic concept of differentiation therapy is that specific identifiable cell signaling pathways maintain stemness in cancer stem cells.[2–6] If the stemness signaling pathways that regulate cancer stem cells can be modified, then the cancer stem cells should divide by symmetric division to produce two daughter cells that progress to cancer transit-amplifying cells. As cancer transit-amplifying cells, both daughter cells would be susceptible to other forms of therapy, including differentiation therapy. The efficacy of differentiation therapy of cancer stem cells was first demonstrated for teratocarcinomas.

TERATOCARCINOMAS

Teratocarcinoma is a malignant cancer that arises from the germinal stem cells, usually in the testes or the ovaries of a young adult.[7] This cancer usually grows rapidly but responds well to therapy, as exemplified by the case of Lance Armstrong. Studies beginning in the 1950s on teratocarcinoma proved not only that these cancers arise from stem cells, but also, more generally, that teratocarcinoma growth is maintained by stem cells, and that therapy can be directed against these cancer stem cells.[8] Usually, fewer than 5% of the cells that make up a teratocarcinoma can be considered cancer stem cells. The malignant teratocarcinoma cells form structures resembling a presomite embryo, the so-called embryoid bodies (*les boutons embryonnaires*),[9] and have characteristics of an embryonic cell (embryonal carcinoma).[10–12] Although the proliferating cells of the embryoid body are undifferentiated, many of the daughter cells of these cells differentiate into mature benign cells[13] so that most of the mass of the tumor consists of differentiated cells. The embryonal carcinoma stem cells may undergo asymmetric division, giving rise to daughter cells that differentiate to form any of more than two dozen well-differentiated adult tissues, including brain, muscle, bone, teeth, bone marrow, eyes, secretory glands, skin, and intestine as well as placenta and yolk sac.[12,14] The ability of retinoids to induce differentiation of teratocarcinoma stem cells proves the principle that it is possible to treat cancer stem cells by differentiation therapy.[4]

Retinoic acid and differentiation therapy of teratocarcinomas

Retinoic acid (RA, vitamin A), in particular, all-trans-retinoic acid (ATRA),[15] or RA in combination with dibutyryl-cAMP, induces differentiation of undifferentiated, immortalized embryonal carcinoma cells into a differentiated epithelial cell type, resembling extraembryonic endoderm.[16,17] RA acts through specific RA nuclear receptors (RARs) to activate c-Fos/c-Jun–mediated transcription[18,19] of genes coding for differentiation products.[20–22] This results in differentiation and decreased proliferation of embryonal carcinoma cells.[23–25] Unfortunately, RA treatment of humans with teratocarcinoma does produce consistent or prolonged clinical improvement.[26] Teratocarcinomas are treated by surgery, followed up by irradiation and/or cisplatin-based cytotoxic chemotherapy,[27] if the surgical removal is incomplete. Fortunately, because teratocarcinomas usually contain yolk sac elements (which produce alpha-fetoprotein) and placental elements (which produce chorionic gonadotropin), these markers can be used to follow the course of treatment very accurately.[28,29] Differentiation therapy has been most successfully applied to human leukemia by reversing the blocks in maturation

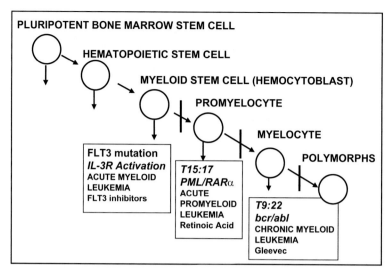

Figure 5–2: Gene translocations, levels of maturation arrest, and differentiation therapy of selected human leukemias. Specific gene translocations lead to expression of signaling molecules that constitutively activate cells at various stages of differentiation in the myeloid lineage: chronic myeloid leukemia (CML; T9:22, Philadelphia chromosome; bcr-abl); acute promyeloid leukemia (APL; T15:17, promyelocytic protein/retinoic acid receptor [PML/RARα] fusion product; see Figure 5–3); acute myeloid leukemia (AML; multiple possibilities using including both an activation signal, such as activation of the interleukin 3 [IL-3] receptor, and a block in apoptosis; see Figure 5–4). CML is effectively treated by Gleevec, which specifically blocks the bcr-abl tyrosine kinase. APL is treated by retinoic acid, which reacts with the retinoic acid in the fusion product and allows the affected cells to differentiate. Treatment of AML by differentiation therapy is still in the experimental stage. Modified from Sell.[6]

responsible for constitutive activation of the leukemia transit-amplifying cells.[6]

LEUKEMIA

Leukemia is a malignant cancer of the blood and lymphoid system. Because of a block in maturation of the leukemia transit-amplifying cells, there is a massive increase in immature white blood cells in the circulation, which changes the color of the blood from red to creamy white; *leukemia* means "white blood" (*leukos*, "white"; *haima*, "blood"). Patients with leukemia usually die because the leukemia cells replace normal immune cells, and inflammatory cells are unable to fight off infections.

Untreated myeloid leukemia is clinically classified on the basis of how rapidly the disease progresses into acute, subacute, or chronic, although there are many variations. Figure 5–2 shows the stages of maturation arrest of chronic myeloid leukemia (CML), acute promyeloid leukemia (APL),and acute myeloid leukemia (AML). CML is arrested at the myelocyte level, APL at the promyelocyte level, and AML at the myeloid progenitor cell level. At each of these stages of maturation, there is proliferation of the leukemia transit-amplifying cells because of gene rearrangements that result in constitutive activation of specific signaling pathways.[6] Only four of the many translocations identified in myeloid leukemia[30] will be discussed to illustrate the principle of differentiation therapy: (1) the t9:22 bcr/abl translocation (Philadelphia chromosome) in CML, which results in constitutive activation of tyrosine phosphorylase[31]; (2) the t15:17 PML/RARα translocation in APL; and (3) two of many possible translocations in AML: t12:13 (FLT3; IL-3R), which activates kinases, and 13q12 ITD FLT3, which blocks apoptosis. The stage of maturation arrest for each of these leukemias is determined by the level at which the transgene product acts (Figure 5–2).

Differentiation therapy of leukemia

Specific differentiation therapy can be directed toward the products of the genetic lesions responsible for the maturation arrest of the leukemic transit-amplifying cells.[6,32]

Chronic myeloid leukemia

The active tyrosine kinase responsible for constitutive activation of myelocytes in Chronic myeloid leukemia (CML) can be effectively blocked by imatinib (Gleevec, manufactured by Novartis; formerly called ST1571)[33,34] and related compounds.[35] Once the constitutive signaling pathway is inhibited, the leukemic transit-amplifying cells differentiate and die by apoptosis.

For almost 50 years, it has been known that most CMLs are associated with a distinct chromosomal abnormality, the Philadelphia chromosome.[36,37] The Philadelphia chromosome is formed by a gene rearrangement that places the breakpoint cluster regions (BCR) of one chromosome next to the Abelson oncogene (*abl*) in another chromosome, resulting in production of a fusion protein (BCR-ABL). The BCR-ABL fusion product is a tyrosine kinase.[38] Increased expression of this tyrosine kinase results in maturation arrest of myeloid cells at the myelocyte stage of differentiation,[37,39] due to activation of downstream signaling molecules such as phosphoinositide 3-kinase.[40] Delineation of the structure of the fusion tyrosine kinase responsible for CML led to the development of a specific small molecule inhibitor, imatinib mesylate (Gleevec, or STI571).[41,42] This inhibitor binds to the active site of fusion protein and prevents binding of ATP and ADP. This prevents the kinase-mediated phosphorylation of the substrate and effectively inactivates the activation pathway. When this signaling pathway is inhibited, the CML cells complete differentiation and die in a few days, like normal mature granulocytes.

Imatinib treatment of patients with CML produces an 80% complete cytogenetic response (CCR) in newly diagnosed patients and a 40% CCR in patients who have relapsed after failure of other therapies.[43] However, residual disease can still be detected by reverse transcriptase–polymerase chain reaction in the bone marrow of patients with CCR. To prevent redevelopment of CML, imatinib treatment must be maintained, or the leukemic stem cells will proliferate and

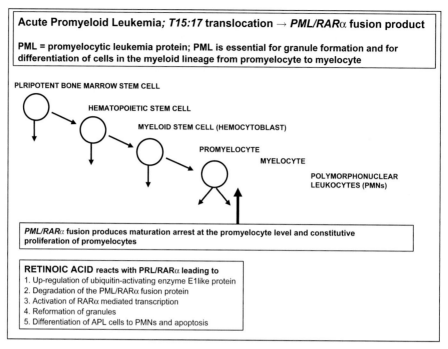

Figure 5–3: Retinoic acid (RA) treatment of APL. RA reacts with the PML-RARα fusion protein and upregulates ubiquitin-activating enzyme E1 like protein, and this causes degradation of the fusion protein. The PML/RARα fusion protein of APL blocks the effect of promyelocytic leukemia protein (PML), which normally is responsible for formation of the nuclear body essential for granule formation and for differentiation of promyelocytes to myelocytes. When PML is inactivated, maturation arrest occurs at the level of the promyelocyte, with accumulation of the cells of APL. Treatment with RA also activates RARα-mediated transcription, which allows reformation of granules, and differentiation of APL cells to terminally differentiated polymorphonuclear leukocytes.

the leukemia will recur.[44] Thus imatinib acts at the myelocitic stage and does not eliminate the leukemic stem cells. Most patients who show a CCR and who are maintained on imatinib remain stable, but some develop mutations in the kinase domain of the *bcr-abl* gene, resulting in a modified fusion product that is no longer inhibited by imatinib. Thus, to cure CML, a way must be found to combine imatinib differentiation therapy with genetic therapy, to block stem cells from reentering the cell cycle and producing a new population of proliferating cells.

Acute promyeloid leukemia

Treatment with RA induces differentiation of acute promyeloid leukemia (APL) cells. RA reacts with the fusion gene product and causes its degradation[45,46] (Figure 5–3). APL is due to a maturation block of myeloid transit-amplifying precursor cells of the myelocytic lineage that express markers of the M3 type (promyelocytes). The maturation arrest is due to formation of a fusion protein between the promyelocytic leukemia protein (PML) and a nuclear receptor for retinoic acid (PML/RARα), the product of a t(15:17) gene rearrangement.[47–49] Normally, PML is found in a discrete, circumscribed nuclear

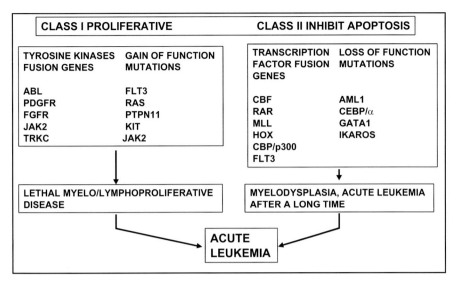

Figure 5–4: Some possible genes activated or lost in AML. AML is usually associated with two genetic changes: one leading to activation of proliferation and the other to loss of apoptosis. One genetic change may result in a chronic myeloproliferative disease. However, where a proliferative lesion (Class I) is combined with a loss of apoptosis lesion (Class II), the chronic disease progresses to AML. The multiple possibilities of gene translocation and gain and loss of function mutations complicate the ability to develop blocking and activating agents for the high number of possible genetic lesions. Figure based on the work of Sell.[5]

structure called a nuclear body.[47,48] In the absence of RA, the fusion product PML/RARα functions as a constitutive transcriptional repressor, disrupting formation of nuclear bodies and blocking promyelocytic differentiation.[50,51] ATRA reacts with the RARα in the fusion protein, upregulates ubiquitin-activating enzyme-E1 like protein (UBE1L), and triggers degradation of the PML/RARα fusion protein.[52,53] This reactivates RARα-mediated transcription, allows reformation of the PML nuclear body, and stimulates differentiation and apoptosis of APL cells.[50,53] Treatment with ATRLA produces complete remission in about 90% of patients with newly diagnosed APL and a complete cure in over 70% (usually when combined with chemotherapy).[54–58] This provides a clear example of the efficacy of combined cytotoxic and differentiation therapies.

Acute myeloid leukemia

Attempts have been made to block the signaling pathways of acutemyeloid leukemia (AML) using small molecules. The lesions of AML are much more complex than for CML or APL. AML progresses rapidly because there is usually more than one molecular lesion (Figure 5–4). The lesions fall into two functional classes: Class I, proliferative, and Class II, apoptosis inhibitory.[59] Class I lesions by themselves cause a chronic myeloproliferative disorder, and Class II lesions, myelodysplasia. Progression to acute disease occurs when there is a second mutation so that both a Class I lesion and a Class II lesion are present. The combination of a proliferative lesion with a lesion that leads to loss of apoptosis results in cells that rapidly proliferate and do not die.[59] Because there are at least two lesions in AML, specific differentiation therapy requires that both lesions be treated. Thus, so far, therapy using small molecules to inhibit the signaling pathways of AML has met with limited success since agents for only one of the two signals are available in most cases.

A common mutation in adult AML occurs in one of the RAS family proteins, the 21-kDa guanine-nucleotide binding protein. RAS proteins require several posttranslational steps for activation, including addition of a farnesyl lipid moiety essential for translocation of the RAS protein to the plasma membrane and for activation of the signal transduction pathway. Agents that inhibit farnesyl transferase block this step, prevent activation, and allow the cells to differentiate.[60] Inhibitors of farnesyl transferase have been reported to induce remission in 20% of older patients with poor prognoses,[61] but they have shown only limited effects in phase 2 clinical trials.[62]

Inhibitors of FMS-like receptor tyrosine kinase 3 (FLT3) are being tested in pediatric AML patients in combination with chemotherapy. FLT3 acts through the Src family of tyrosine kinases.[59,63] FLT3 also interacts with a nuclear fusion protein formed as a result of other gene translocations such as 11q23, the mixed-lineage leukemia gene.[64] The formation of heterodimers between these proteins transforms hematopoietic precursors in vitro and may be a critical signal for maturation arrest in AML types M4 and M5. Approximately 30% of AML patients have an activating mutation of FLT3.[65] Small-molecule inhibitors of FLT3 selectively kill transformed cells that have activating mutations of FLT3. In clinical trials, several FLT3 inhibitors have induced biological responses, manifested as large reductions in numbers of peripheral blood leukemic cells, but complete remissions are very rare and the biological effects are of limited duration.[65] The effect of these inhibitors in AML is not as great as the effect of Gleevec in CML. Since development of CML may depend solely on activation of *bcr-abl*, and since AML usually involves both an activation of proliferation and inhibition of apotosis,[59,66] the combination of FLT3 inhibitors with other drugs should be more effective than the use of FLT3 inhibitors alone. Combinations of FLT3 inhibitors with other modalities (azacitidine, chemotherapy, bone marrow transplantation) or with small-molecule inhibitors of other leukemia genes are an active field of investigation. Radiation treatment or chemotherapy could first be used to kill proliferating cells, followed by targeted therapy to prevent proliferation of cancer stem cells and reformation of the leukemia, after conventional therapy is discontinued.

Regrowth of leukemias

The leukemia transit-amplifying cells are the target for these examples of differentiation therapy. However, the molecular lesions are present in all the cells of the myeloid lineage, including the myeloid stem cells. Differentiation therapy removes the block at the transit-amplifying cell level, but when therapy is discontinued, the leukemia will reform from leukemia stem cells. The tumor stem cells are not affected by the therapy so that when differentiation therapy is discontinued, the leukemia transit-amplifying cells arising from the tumor stem cell will no longer be inhibited and will reform the leukemia.[6,32]

RADIATION AND DRUG RESISTANCE OF STEM CELLS

Resting stem cells of both normal and cancerous tissues are generally much more resistant to radiation treatment and chemotherapy than are the proliferating transit-amplifying cells.[67] To be curative, local irradiation of cancer must completely eliminate not only the transit-amplifying cells, but also all the cancer stem cells. On the basis of the clinical observation of "accelerated repopulation," radiation therapists have concluded that cancers contain stem cells that are resistant to radiation therapy.[68] Accelerated repopulation refers to the more rapid regrowth of the cancer during treatment gaps. The probable culprits are stem cells that are in G_0 (resting phase) at the time when therapy is administered. Accelerated repopulation is most likely explained by activation of the cancer stem cells to produce more stem cells that exhibit symmetric division during treatment gaps so that regrowth of the cancer during intervals between treatments occurs more rapidly. The key to more effective therapy may be to find a way to convert both daughter cells of the cancer stem cell to a non-stem-cell phenotype.

Normal and cancer stem cells are also both resistant to chemotherapy.[3] Stem cells express high levels of antiapoptotic proteins[69,70] and multidrug transporter molecules.[71,72] Stem cells are not only able to pump out cytotoxic drugs, but also eliminate the fluorescent dye Hoechst 33342.[72,73] This characteristic allows us to sort out putative stem cells, the so-called side population (SP) cells, which do not take up the dye, by flow cytometry. In breast cancer stem cells, the ATP binding cassette (ABC) transporter ABCG2, also known as breast cancer resistance protein (BCRP1), can also contribute to dye exclusion typical of the SP cells. These properties of cancer stem cells explain why it has been difficult to develop an effective cancer therapy based primarily on attack directed at proliferating cells. Although antimetabolic drugs or radiation may be effective against these tumor transit-amplifying cells, they clearly cannot be expected to affect the cancer stem cell. Thus it is necessary to design a cancer therapy that will not only be effective against the proliferating cancer cells, but will also block the nonproliferating cancer stem cells. The principle of regrowth of treated cancers from resistant stem cells is illustrated by treatment of leukemia.

Leukemic stem cells and response to therapy

Cure of leukemia requires elimination of the most primitive leukemic stem cells, which carry the leukemic translocation or mutation, as well as the proliferating leukemic transit-amplifying cells.[74] The resistance of cancer stem cells to chemo- and radiation therapy is clearly illustrated by the response of AML to cyclic chemotherapy. When AML is first detected, the tumor load is in the range of 10^{12} cells. In general, chemo- or radiation therapy will be effective in eliminating

99.9% of the AML cells.[32] This kill percentage is consistent either with the idea that less than 1 in 1,000 of the AML cells is a stem cell that is resistant to the therapy or with the idea that treatment is ineffective against a leukemic cell that is not dividing when the therapy is administered. Since chemo- and radiation therapies are directed against proliferating cells (the growth fraction or transit-amplifying cells) of the AML, the stem or resting tumor cell is not affected by the therapy. The frequency of the therapy-resistant cell appears to be somewhat higher than the frequency of tumor-initiating stem cells that was determined via transplantation,[75,76] suggesting that some of the chemoresistant cells are not true leukemic stem cells.[77] In any case, when the therapy is discontinued to allow the normal hematopoietic cells to recover, the AML stem cells will also begin to regrow and will produce more leukemic transit-amplifying cells. Then, a second cycle of therapy will be given that will once again eliminate 99.9% of the leukemic cells. Since the putative AML stem cells will again be resistant, the tumor will again regrow. After four cycles of therapy, some leukemias will be cured, suggesting that, in some leukemias, the most primitive bone marrow stem cell is not mutated.[32] On the other hand, the genetic change in many AMLs is present in the most primitive stem cells, which are resistant to chemotherapy. At this point, the curative therapy modality must be changed to ablation by irradiation or chemotherapy and restoration of myelopoiesis by bone marrow transplantation.

Cancer stem cells or quiescent (dormant) cancer cells

The time-honored concept that cancers contain stem cells responsible for the capacity to transplant the tumor (tumor initiation) and for resistance to therapy and regrowth of cancers after what appears to be successful therapy has recently been rediscovered.[3,78,79] However, it is not clear if cancers actually contain stem cells in the same manner as normal tissues or if some cells of a cancer are dormant at the time of therapy and are not susceptible to treatments that are directed to the cancer transit-amplifying cells. For example, these alternative explanations were addressed by Trott[80] and by Denekamp[81] in 1994. Trott[80] concluded that a proportion of cells ranging from 0.1% to 100% of all cells from transplantable mouse tumors met the criteria of a tumor stem cell, that is, "regrowth of the tumour preceded by clonal expansion from a single cell with unlimited proliferative potential." He concluded that tumors contain stem cells similar to the lineage of cells taking part in normal tissue renewal; in other words, cancers contain the same populations of cells as normal tissue. This interpretation was the same as that derived earlier from the study of the cells of teratocarcinoma.[82] On the other hand, Denekamp,[81] considering the same evidence, deduced that the putative cancer stem cells are merely the least differentiated cells in the cancer population and only appear functionally and kinetically different from the mass of tumor cells. She concluded that the cancer stem cell is not as clearly definable as the normal-tissue stem cell and that cells in a cancer that are resistant to therapy may be cancer cells that are quiescent when the therapy is administered.

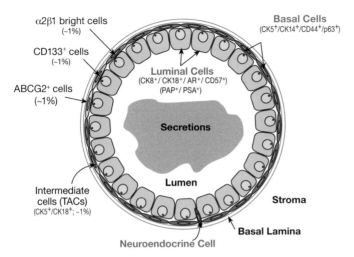

Figure 2–1: Prostatic glands are ductal structures composed of phenotypically and functionally distinct cells.

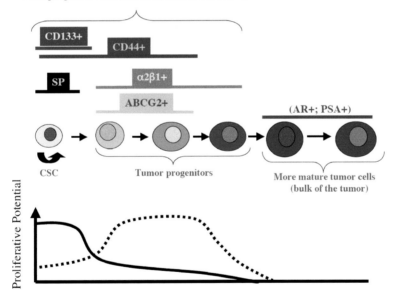

Figure 2–3: A tumorigenic hierarchy provides evidence of prostatic cancer stem cells (CSCs) in xenograft tumors.

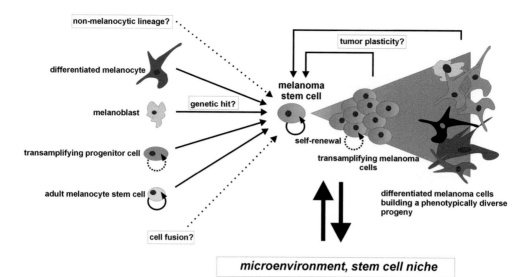

Figure 3–2: Melanoma stem cell hypothesis. Recent data indicate that there is a subpopulation of cells within melanoma that are characterized by stem cell features such as indefinite self-renewal and differentiation capacity.

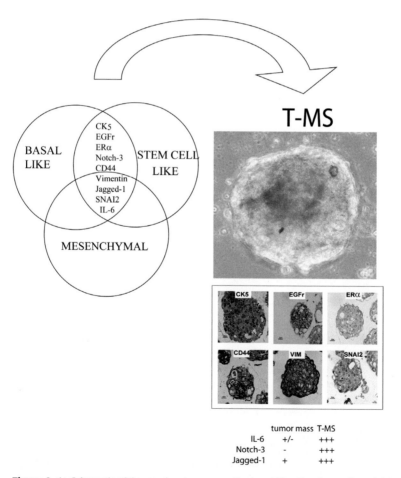

	tumor mass	T-MS
IL-6	+/-	+++
Notch-3	-	+++
Jagged-1	+	+++

Figure 4–1: Schematic of the overlapping among the basal like, the stem cell, and the mesenchymal phenotype and its representation in human ductal breast carcinoma–derived (T-) MS.

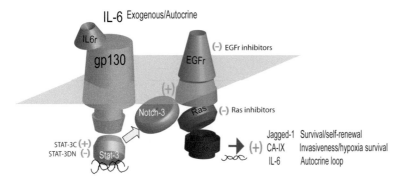

Figure 4–2: Schematic of IL-6–controlled activities in MS and breast cancer cells.

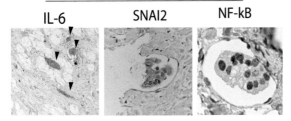

Figure 4–3: Immunohistochemical analysis of lymphovascular tumor emboli.

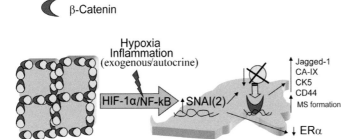

Figure 4–4: Schematic of the relationship between the hypoxia/inflammation-dependent upregulation of SNAI gene family members and the basal/stem cell–like phenotype.

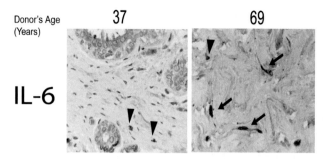

Donor's Age
(Years)

37   69

IL-6

Figure 4–5: Immunohistochemical analysis of normal mammary gland stromal cells (*arrowhead,* macrophages; *arrows,* fibroblasts).

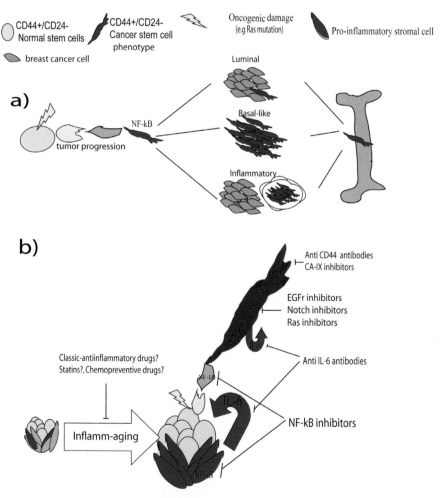

Figure 4–6: (a) Schematic of the basal/stem cell–like phenotype in various breast cancer subtypes. (b) Contribution of an aged/pro-inflammatory stromal cell environment.

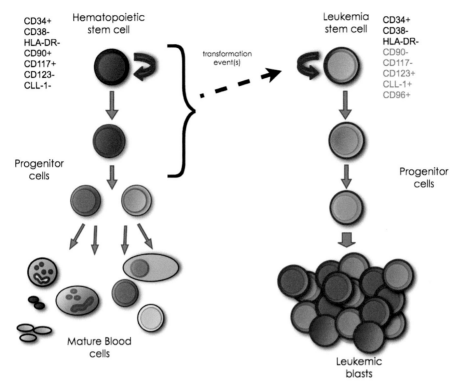

Figure 6–1: Leukemia stem cells (LSCs) are thought to arise from a normal hematopoietic stem cell or progenitor cell that undergoes a malignant transformation event.

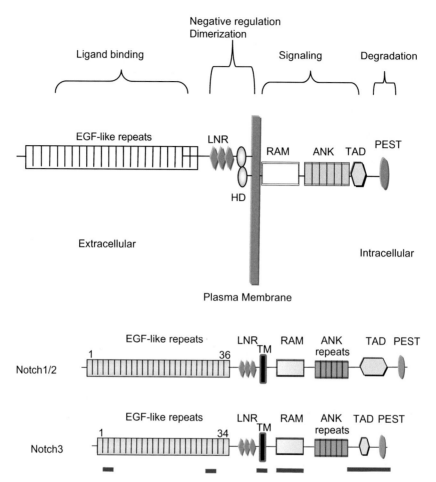

Figure 8–1: Diagram of Notch receptor.

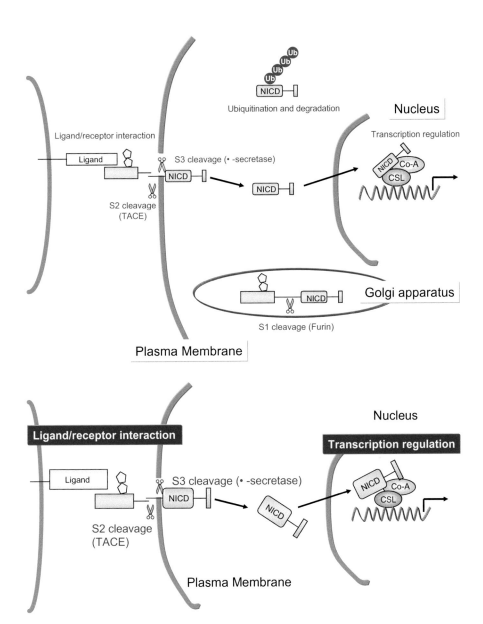

Figure 8–2: Notch pathway elements.

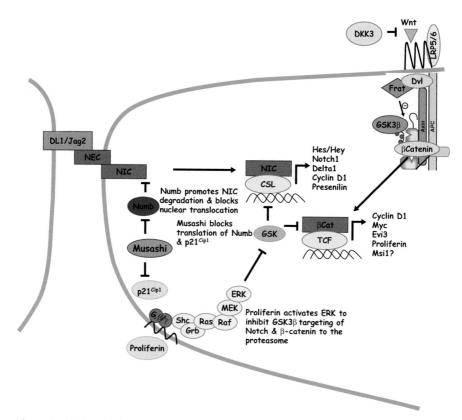

Figure 9–1: Musashi signaling pathways associated with mammary stem cells.

These alternatives have more recently been reconsidered and updated.[83] Kern and Shibata,[84] using a mathematical analysis, point out that tumor-initiating capacity could be a varying probabilistic potential for all tumor cells, rather than a quantal and deterministic feature of a minority of tumor cells. Identification of tumor-initiating cell populations through the use of marker phenotypes could preferentially enrich for cells able to transplant tumors, but even with the best purification systems, the so-called nontumorigenic cell population will contain up to 3% of tumorigenic cells.[85] Are these cancer stem cells or transit-amplifying cells in the cancer cell lineage arrested at an earlier stage of differentiation than the majority of the nontumorigenic population? The recent attempts to separate cancer stem cells from non–stem cells using flow-cytometric separations depend on cell surface markers that may change expression or be masked by cell surface carbohydrates. Thus the fractionation procedure itself may change the expression of the marker or alter the phenotype of the cell. In addition, the significance of transplantation of human cancer cells into severe combined immunodeficiency disease (SCID) mice as an indicator of the tumor-initiating property of cancer stem cells has come into question.[77] In contrast to the finding that only 1 in 250,000 human leukemic cells is transplantable into SCID mice, essentially all of the cells of a mouse B-cell lymphoma will produce tumors when injected into nonirradiated congenic recipients,[77] previously reported in 1937.[86] Thus it is possible that the tissue microenvironment of a SCID mouse limits the ability of the human leukemic cells to form a tumor. If so, the low fraction of cells in human leukemia transplantable to SCID mice could be due to an incompatible microenvironment,[77] not to a property of the transplanted cells.

Regardless, if the explanation for tumor initiation and resistance to therapy is because of cancer stem cells, or the presence of dormant cancer cells, it is clear that efforts to develop better therapeutic approaches need to be directed to these cells. In fact, the properties of so-called cancer stem cells and dormant cancer cells may be similar. Therapies designed to alter cancer stem cells may also affect dormant cancer cells. If so, it is possible that new therapies can be directed to some of the shared characteristics of the cancer stem cells and dormant cancer cells[3–5] such as the signals that maintain stemness.[87,88]

CANCER STEM CELL–DIRECTED THERAPY

Regardless of whether cancers are maintained by stem cells or by a population of cells that are in G_0 at the time of treatment, the signaling pathways of the treatment-resistant cells can be exploited in an effort to eliminate the therapy-resistant cells. The rationale for this approach is presented in Figure 5–5.[5] It is based on the supposition that cancer stem cells or dormant cancer cells are maintained as stem cells or dormant cells by specific stem cell signaling pathways[87] and that blocking of the stemness pathways will force cancer stem cells to become transit-amplifying cells. As transit-amplifying cells, the cancer will lose resistance to other forms of therapy.

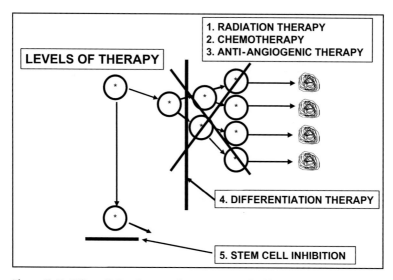

Figure 5–5: Differentiation therapy of cancer stem cells. Chemotherapy, radiotherapy, and antiangiogenic therapy are directed to the actively proliferating transit-amplifying cells of a cancer. When these therapies are discontinued, the cancer regrows from the therapy-resistant cancer stem cells. Differentiation therapy blocks the activation signals, causing maturation arrest. However, when differentiation therapy is discontinued, the cancer will reform from the cancer progenitor cells. Stem cell inhibition is directed against the signals that keep a cancer stem cell a stem cell. By blocking or reversing the stemness signals, it may be possible to force the cancer stem cell to differentiate. Modified from Sell.[5]

Inhibition of cancer stem cell signaling pathways

A number of signaling pathways have been identified in normal[89] and cancer[90–92] stem cells. Inhibition of normal stem cells by RNA interference allows the stem cells to begin differentiation.[89] Using loss-of-function strategies,[93] knockdown of Oct-4[94,95] or Nanog[96] promotes differentiation of human embryonal stem cells, as determined by changes in morphology, growth rates, gene profiling, and phenotype. Thus the principle that the stem cell properties of normal stem cells are maintained by stemness signals and that blocking these signals induces differentiation has been established. In normal asymmetric stem cell division, the signals that determine which daughter cell remains a stem cell, and which begins the process of determination, may also be the signals that control the growth and differentiation of cancer cells.[97]

Cancer stem cell signals

Cancer stem cell signals include Oct-4, Wnt/β-catenin, Notch, BMP (bone morphogenetic protein), Janus family kinase (JFK), sonic hedgehog (shh), and others.[89,98–104] Some potential targets for therapy are listed in Table 5–1. Increased activities of Oct-4, Wnt, and shh have been specifically identified in breast cancer.[99,101–103] The Wnt/β-catenin family has been consistently associated with cancer cells of various types.[103–105] The principle of how to direct targeted therapy

Table 5–1. Some potential targets for iRNA inhibition in cancer therapy

Cancer type	Targets
Breast	Oct-4, Wnt, Shh, Her2/neu, EGFR, epithelial specific antigen (ESA), CD44, Notch, transforming growth factor-beta (TGF-β), BMI1
Leukemia/non-Hodgkin's lymphoma	CD33, CD45, etc.
Cancer stem cells	Bmi1, c-kit, Notch 1, CD-133, chemokine receptor CXCR4, CD34, SCA-1, Thy-1, EED, Oct4, Lmo4
Solid tumors	Oct-4, Wnt, SLAM family members CD48, CD150, and CD244

against cancer stem cells will be addressed using the example of breast cancer; the conclusions also apply to other solid cancers.

Breast cancer

A number of signaling pathways are active in breast cancer cells. For the purpose of this chapter, we will concentrate on three pathways: Wnt, Her-2/Neu, and BMI1.

Wnt

The Wnt pathway is a prime target for gene therapy of breast cancer stem cells. Wnt-1 and Wnt-6 have been found to be highly expressed in both normal and malignant breast tissues, but Wnt-7b is downregulated in breast cancer.[105] Wnt-1 mRNA is upregulated by β-estradiol in MCF-7 breast cancer cells,[106] as is Wnt-5B, but Wnt5A is not.[107] Wnt-3 and Wnt-3a mRNAs are coexpressed in MCF-7 cells and are downregulated together by β-estradiol in MCF-7 cells.[108] Transgenic expression of Wnt-1 or Wnt-10b in mouse mammary gland leads to lobuloalveolar hyperplasia, with progression to cancer.[103] Wnt family members associated with the Wnt/β-catenin pathway are expressed in breast cancer stem cells, and Wnt family members associated with the Wnt/Ca^{++} pathways are expressed in non–stem cells.[98] Wnt signaling is increased in other stem cell cancer systems such as gastrointestinal cancer.[109]

Her-2/Neu

Another prominent stem cell signaling pathway in breast cancer involves upregulation of the epidermal growth factor receptor (EGFR) in the form of the proto-oncogene Her-2/neu (c-erb-B2). Her-2/neu acts through Wnt-mediated β-catenin signaling and loss of E-cadherin–mediated cell adhesion.[110] Loss of E-cadherin and disruption of cell–cell adhesion are associated with progression and metastasis of breast cancer.[111] Alterations of E-cadherin and caveolin1 protein activate the nuclear translocation of β-catenin, thereby activating the β-catenin signaling pathway. The natural action of EGF is to increase levels of β-catenin and expression of c-MYC, which is a target gene of the β-catenin-TCF/LEF1 pathway.[112] Activation of Her-2/neu and β-catenin appears to be critical in the progression of breast cancer to metastasis. Blockade of these signals may be an effective means of inhibiting breast cancer stem cells and treating metastatic breast cancer.[113] Thus it is proposed that gene therapy of cancer should be directed toward blocking the activation pathways discussed previously.[90]

BMI1

Recent clinical genomics data provide compelling evidence for the role of BMI1-activated oncogene expression in breast and other cancers. These data also provide powerful evidence supporting a cancer stem cell hypothesis and suggest that gene expression signatures associated with the stemness state of a cell (defined as phenotypes of self-renewal, asymmetrical division, and pluripotency) might be predictive of cancer therapy outcome.[114] A mouse/human comparative cross-species translational genomics approach was utilized to identify an 11-gene signature that distinguishes stem cells with normal self-renewal function from stem cells with drastically diminished self-renewal ability due to the loss of the *BMI1* oncogene as well as consistently displaying a normal stem cell–like expression profile in distant metastatic lesions, as revealed by the analysis of metastases and primary tumors in both a transgenic mouse model of prostate cancer and cancer patients.[114] Kaplan–Meier analysis confirmed that a stem cell–like expression profile of the 11-gene signature in primary tumors is a consistent powerful predictor of a short interval to disease recurrence, distant metastasis, and death after therapy in cancer patients diagnosed with 12 distinct types of cancer.[114–117] Most prominent is the presence of a conserved *BMI1* oncogene-driven pathway, which is similarly activated in both normal stem cells and a clinically lethal therapy-resistant subset of human tumors diagnosed in a wide range of organs and uniformly exhibiting a marked propensity toward metastatic dissemination. Consistent with this idea, the essential role of the *BMI1* oncogene activation in prostate cancer metastasis as well as in the maintenance of a self-renewal ability and high malignant potential of human breast cancer stem cells has been demonstrated.[114,118,119] In preliminary experiments in our laboratory, it was shown that transfection of breast cancer cells from transgenic mice with RNAi to *BMI1* greatly reduces the ability of the breast cancer cells to initiate tumors on transplantation (G. Glinsky et al., unpublished data).

Other cancers

It is not possible to list all of the potential pathways in this brief chapter. Signals that have been specifically identified as being associated with cancer stem cells and poor prognosis are Bmi1, Oct4 (Pou5F1), EED, and Lmo4.[116] It is likely to be more effective to attempt to inhibit these signals, rather than signals such as Notch and NF-κB, which are more highly expressed in normal stem cells than in cancer stem cells.[92,120] Although some of the signal proteins, such as Notch and NF-κB, may eventually be targeted, it would appear more effective to target signals that are more prominent in cancer cells than in normal stem cells. In any case, if this approach is ever to be used, it is necessary to identify specific inhibitors for cancer stem cell signals and methods or approaches to deliver the inhibitor effectively in vivo.

CANCER STEM CELL INHIBITION

Possible inhibitors of cancer stem cell signaling pathways include iRNA and specific small-molecule inhibitors of the signaling pathway.[5,90]

Inhibitory RNA

RNA interference (RNAi) has been proposed for inactivating cancer stem cells.[94–96,121] An oligonucleotide sequence complementary to messenger RNA (mRNA) will target the nucleic acid in the cell and lead to degradation of the mRNA. RNAi can inhibit gene expression both in vitro and in vivo.[122–124] The demonstration that an oligonucleotide sequence complementary to the 3' end of Rous sarcoma virus was able to block viral replication in chicken fibroblasts provided a proof of principle for the use of RNAi to block cancer cells.[125] If RNAi can be used to block the signals that maintain stem cell characteristics,[89] it may be possible to block cancer stem cells by treatment using siRNA sequences specific for Nanog, Oct-4, Sox2, selected Wnt family members, Notch-4, Her-2/neu, β-catenin, shh, and BMI1, either singly or in combination.[90] The problems associated with gene therapy of cancer in general also apply to gene therapy of cancer stem cells. These are obtaining the appropriate ologonucleotide and effective delivery mechanisms for inhibition.

Oligonucleotides

Advances in automated DNA synthesis and chemical modifications that increase resistance to nuclease digestion have made it much easier to design and make specific siRNA sequences that inhibit stem cell signals.[126] Inhibition of cancer stem cells following killing of cancer transit-amplifying cells may provide a one-two punch to cancer cells. In a proof-of-principle experiment, it was shown that treatment of a mixture of chronic myeloid leukemic cells and normal hematopoietic progenitor cells in vitro with *bcr-abl* antisense oligonucleotides combined with mafosfamide was highly effective in killing leukemic progenitor cells, yet spared a much higher number of normal progenitor cells than did high-dose mafosfamide treatment alone.[127]

Delivery of iRNA

Whereas inhibition by iRNA has been shown in some instances in vitro, major problems exist in adapting iRNA inhibition for use in vivo.[123,124] First, the iRNA must be injectable in a form that is not degraded and at least has a chance to reach tumor cells intact. Proposed injectable forms include free, cyclodextrin polymer-conjugated, carbohydrate-modified, liposomal nanopeptide carrier, biologic nanopeptide vehicle (ENGeneICDeliver Vehicle), and lentivirus or adenovirus constructs. Obviously, such a long list implies that no single approach has worked well, and much needs to be done to devise a successful and general mode of delivery. Also, the route of injection may be critical. For example, free iRNA would not be expected to survive to reach the cancer unless it is injected directly into the tumor. The various routes of delivery proposed, besides direct injection into the tumor, include systemic, nasal, intraperitoneal, and intrahepatic. Nanoimmunoliposome complexes of iRNA against HER-2 mRNA, encapsulated by a cationic liposome and decorated with anti-TfR single-chain antibody fragments, have been targeted to primary and metastatic lesions in SCID mice transplanted with human breast and pancreatic cancers.[124] Problems arise in how to deliver genes or gene products to cancer stem cells, and in how to avoid affecting normal tissue stem cells. Liposomal carriers and virus-based expression vectors may be used to attempt to deliver concentrations of siRNAs that will be therapeutic.[128] Lentiviral vectors appear to be the best for gene delivery because they infect both noncycling and postmitotic cells, and the transgenes expressed from lentivirus are not silenced during differentiation of embryonic stem cells.[129,130]

It may be possible to direct siRNA sequences specifically to target cancer stem cells via receptors that are known to be expressed on stem cells. In this approach, any of a number of receptors on normal stem cells could be overexpressed on cancer stem cells. Targeting steroid receptors on cancer cells has been proposed for cancer gene therapy.[128] Other possibilities include the epithelial-specific antigen (ESA) or CD44 phenotype described for breast cancer stem cells.[131–133] Constructions combining an antibody to ESA or CD44 with siRNA against stem cell signals might allow preferential delivery and inhibition of breast cancer stem cells. Other possible cancer cell receptors or markers may be addressed in this way. Some of these are c-kit,[134,135] estrogen receptor and EGFR,[133,136] CD-133,[137] and chemokine receptor CXCR4.[138] Other markers present on hematopoietic stem cells that might also be expressed on cancer stem cells of other tissues include CD34, SCA-1, and Thy-1[139] as well as SLAM family members CD48, CD150, and CD244,[140] which may be differentially expressed depending on the degree of differentiation of the cancer stem cells.

The potential sharing of "ondodevelopmental markers"[141] between cancer stem cells and normal tissue stem cells raises the issue of delivery of inhibitory molecules to the wrong cells, in particular, delivery to hematopoietic or gastrointestinal (GI) stem cells. The resting stem cells in these organs are required for regeneration of the blood or GI transit-amplifying cells, which are highly susceptible to radiation treatment or chemotherapy due to their high turnover rate.[67,68] To realize application of gene therapy to cancer stem cells, it may be necessary to screen cancer stem cells for unique markers that might not be shared with normal tissue stem cells.[136,142] It may also be possible to limit application to local areas so as to avoid systemic therapy or else devise procedures to protect vulnerable

Table 5–2. Some inhibitors of major stem cell signaling pathways[a]

Signaling pathway	Inhibitor
JAK-STAT	APS
Notch	γ-secretase inhibitor (DAPT)
MAPK/ERK	RAF kinase inhibitors/U0126
PI3-K/Akt	Rapamycin (LY294002)
NF-κB	I- κB, PTDC
Wnt/β-catenin	NSAID, GSK-3, sFRPs, DKK, Axin
TGFβ (BMP)	SMAD6,7; Lefty1,2; Gremlin, SM16, etc.
Sonic hedgehog (shh)	Cyclopamine
Oct-4/Sox2/Nanog	Tcf3

[a] Data adapted from Sell.[87]

normal tissue-renewing cells (such as by combining cancer stem cell gene therapy with erythropoiesis-stimulating agents).[143]

Molecular inhibitors

Molecular inhibitors for major stem cell signaling pathways are listed in Table 5–2, and the full names of these inhibitors appear in Table 5–3. Again, most of these agents have been shown to have some inhibitory effect on cancer cells in vitro, but how this information can be applied in vivo, and whether or not the agents will have deleterious effects on normal tissue stem cells, remains to be determined. Much of the data now being collected on the use and efficacy of

Table 5–3. Some molecular inhibitors of signaling pathways[a]

Inhibitor	Full name of inhibitor
APS	Adaptor molecule (pleckstrin homology and SH-2 domains)
NSAID	Nonsteroidal anti-inflammatory drugs
GSK-3	Glycogen synthesis kinase-3
sFRPs	Secreted Frizzled-related proteins
DKK	Dickkopf family (WIF-1, Cerebus)
DAPT	γ-secretase inhibitor, *N*-[N-(3,5-difluorophenacetyl)-L-alanyl]-S-phenylglycine t-butyl ester
SMAD6,7	Related to *Drosophila* Mad (mothers against decapentaplegic), inhibitor of SMAD transcription factors for TGF-β pathway
Lefty1,2	Inhibitor of Activin activation of TGF-β pathway
Gremlin	Inhibitor of BMP activation of TGF-β pathway
LY294002	Selective PI3 kinase (PI3K) inhibitor
I-κB	Inhibitor of κB
PTDC	Sodium pyrrolidinethiocarbamate
U0126	MAP kinase inhibitor [1,4-diamino-2,3-dicyano-1,4-bis (o-inophenylmercapto)butadiene ethanolate]
SM16	Small molecular inhibitor of TGF-β type I receptor kinase (ALK5)
Tcf3	Repressor of Wnt target genes
Velcade	Blocks NF-κB

[a] Data adapted from Sell.[87]

molecular inhibitors are restricted by confidentiality agreements of the commercial enterprises supporting the work. Even if some of these approaches eventually do work out, a word of caution is needed because of the tendency of cancer stem cells to mutate and change characteristics, usually to a more malignant form.[144] This property makes the cancer stem cell a moving target for specific therapy.

REFERENCES

1. Pierce GB, Wallace C. Differentiation of malignant to benign cells. Cancer Res 1971;**31**:127–134. AN AACR SCIENTIFIC LANDMARK
2. Sell S, Pierce GB. Biology of disease: maturation arrest of stem cell differentiation is a common pathway for the cellular origin of teratocarcinomas and epithelial cancers. Lab Invest 1994;**70**:6–21. AN AACR SCIENTIFIC LANDMARK
3. Reya T, Morrison SJ, Clarke MF, Weissman IL. Stem cells, cancer, and cancer stem cells. Nature 2001;**414**:105–111.
4. Sell S. Stem cell origin of cancer and differentiation therapy. CR Oncol/Hematol 2004;**51**:1–28.
5. Sell S. Potential gene therapy for cancer stem cells. Curr Gene Ther 2006;**6**:579–591.
6. Sell S. Leukemia: stem cells, maturation arrest and differentiation therapy. Stem Cell Rev 2005;**1**:197–205.
7. O'Hare MJ. Teratomas, neoplasia and differentiation: a biological overview I. The natural history of teratomas. Invest Cell Pathol 1978;**1**:39–63.
8. Pierce GB, Shikes R, Fink LM. *Cancer: a problem of developmental biology*. Englewood Cliffs, NJ; Prentice Hall, 1978.
9. Peyron A. Sur la presence des cellules genitales primordiales dans les boutons embryonnaires des embryomes parthenogenetiques chez l'homme. CR Acad Sci (Paris) 1938;**206**:1680–1683.
10. Pierce GB, Dixon FJ. Testicular teratomas. I. Demonstration of teratogenesis by metamorphosis of multipotential cells. Cancer 1959;**12**:573–583.
11. Pierce GB, Dixon FJ, Varney IL. Teratocarcinogenic and tissue forming potentials of the cell types comprising neoplastic embryoid bodies. Lab Invest 1960;**9**:583–602.
12. Kleinsmith LJ, Pierce GB. Multipotentiality of single embryonal carcinoma cells. Cancer Res 1964;**24**:1544–1551.
13. Pierce GB, Varney EL. An *in vitro* and *in vivo* study of differentiation in teratocarcinomas. Cancer 1961;**14**:1017–1029.
14. O'Hare MJ. Teratomas, neoplasia and differentiation: a biological overview. I. The natural history of teratomas. Invest Cell Pathol 1978;**1**:39–63.
15. Spira AI, Carducci MA. Differentiation therapy. Curr Opin Pharmacol 2003;**3**:338–343.
16. Strickland S, Madavi V. The induction of differentiation in teratocarcinoma stem cells by retinoic acid. Cell 1978;**15**:393–403.
17. Grover A, Adamson ED. Evidence for the existence of an early common biochemical pathway in the differentiation of F9 cells into visceral or parietal endoderm: modulation by cyclic AMP. Dev Biol 1986;**114**:492–503.
18. Wasylyk B, Imler JL, Chatton B, Schatz C, Wasylyk C. Negative and positive factors determine the activity of the polyoma virus enhancer alpha domain in undifferentiated and differentiated cell types. Proc Natl Acad Sci USA 1988;**85**:7952–7956.
19. Oshima RG, Abrams L, Kulesh D. Activation of an intron enhancer within the keratin 18 gene by expression of c-*fos* and c-*jun* in undifferentiated F9 embryonal carcinoma cells. Genes Dev 1990;**4**:835–848.
20. Boylan JF, Lufkin T, Achkar CC, Taneja R, Chambon P, Gudas LJ. Targeted disruption of retinoic acid receptor α (RARα) and RARγ results in receptor-mediated alterations

in retinoic acid-mediated differentiation and retinoic acid metabolism. Mol Cell Biol 1995;**15**:843–851.

21. Choi SK, Yeh J-C, Cho M, Cummings RD. Transcriptions regulation of 1,3-galactosyltransferase in embryonal carcinoma cells by retinoic acid. J Biol Chem 1996;**271**:3238–3246.
22. Congyi C, Gudas LJ. Murine laminin B1 gene regulation during the retinoic acid- and dibutyryl cycic AMP-induced differentiation of embryonic stem cells. J Biol Chem 1996;**271**:6810–6818.
23. Sun SY, Yue P, Mao L, Dawson MI, Shroot B, Lamph WW, Heyman RA, Chandraratna RAS, Shudo K, Hong WK, Lotan R. Identification of receptor-selective retinoids that are potent inhibitors of the growth of human head and neck squamous cell carcinoma cells. Clin Cancer Res 2000;**6**:1563–1573.
24. Sun SY, Lotan R. Retinoids and their receptors in cancer development and chemoprevention. Crit Rev Oncol Hematol 2002;**41**:41–55.
25. Mehta K. Retinoids as regulators of gene transcription. J Biol Regul Homeost Agents 2003;**17**:1–12.
26. Trump DL. Retinoids in bladder, testes and prostate cancer: epidemiologic, preclinical and clinical observations. Leukemia 1994;**8**(Suppl 3):50–54.
27. Yoskitake T, Itoyama S. Treatment of primary mediastinal germ cell tumors. Tinsho Kyobu Geka 1989;**9**:29–34.
28. Bartlett NL, Freiha FS, Torti FM. Serum markers in germ cell neoplasms. Hem/Oncol Clin N Am 1991;**5**:1245–1260.
29. Stenman U-H, Alfthan H. Markers for testicular cancer. In *Tumor markers* (Diamandis E, Fritsche H, Lilha H, Chan D, Schwartz M, eds). Washington, DC; AACC Press, 2002:351–359.
30. Rowley JD. Nonrandom chromosomal abnormalities in hematologic disorders of man. Proc Natl Acad Sci U S A 1975;**72**:152–156.
31. Nowell PC. Diagnostic and prognostic value of chromosome studies in cancer. Ann Clin Lab Sci 1974;**4**:234–240. AN AACR SCIENTIFIC LANDMARK
32. Baird SM. Hepatopoietic stem cells in leukemia and lymphoma. In *Stem cells handbook* (Sell S, ed). Totowa, NJ; Humana Press, 2004:163–175.
33. Drucher BJ, Ralpaz M, Resta DJ, Peng B, Buchdunger E, Ford JM, Lydon NB, Kantarjian J, Capdeville R, Ohno-Jones S, Sawyers CL. Efficacy and safety of a specific inhibitor of the BCR-ABL tyrosine kinase in chronic myeloid leukemia. N Engl J Med 2001;**334**:1031–1037. AN AACR SCIENTIFIC LANDMARK
34. Druker BJ, Guilhot F, O'Brien SG, Gathmann I, Kantarjian H, Gattermann N, Deininger MW, Silver RT, Goldman JM, Stone RM, Cervantes F, Hochhaus A, Powell BL, Gabrilove JL, Rousselot P, Reiffers J, Cornelissen JJ, Hughes T, Agis H, Fischer T, Verhoef G, Shepherd J, Saglio G, Gratwohl A, Nielsen JL, Radich JP, Simonsson B, Taylor K, Baccarani M, So C, Letvak L, Larson RA; IRIS Investigators. Five year follow-up of patients receiving imatinib for chronic myeloid leukemia. N Engl J Med 2006;**355**:2408–2417.
35. Puttini M, Coluccia AM, Boschelli F, Cleris L, Marchesi E, Donella-Deana A, Ahmed S, Redaelli S, Piazza R, Magistroni V, Andreoni F, Scapozza L, Formelli F, Gambacorti-Passerini C. In vitro and in vivo activity of SKI-606, a novel Src-Abl inhibitor, against imatinib-resistant Bcr-Abl+neoplastic cells. Cancer Res 2006;**66**:11314–11322.
36. Nowell PC, Hungerford DA. A minute chromosome in human granulocytic leukemia. Science 1960;**132**:1497–1499.
37. Randolph TR. Chronic meylocytic leukemia – Part I: history, clinical presentation and molecular biology. Clin Lab Sci 2005;**18**:38–48.
38. Tefferi A, Dewald GE, Litsov MO, Cortes J, Mauro MJ, Talpaz M, Kantarjian HM. Chronic myeloid leukemia: current application of cytogenetics and molecular testing for diagnosis and treatment. Mayo Clin Proc 2005;**80**:390–402.
39. O'Brien S, Tefferi A, Valent F. Chronic myelogenous leukemia and myeloproliferative disease. Hematology 2004:146–162.

40. Kharas MG, Fruman, DA. ABL oncogenes and phosphinositide 3-kinase: mechanism of activation and downstream effects. Cancer Res 2005;**65**:2047–2053.

41. Deninger M, Buchdunger E, Druker BJ. The development of imatinib as a therapeutic agent for chronic myeloid leukemia. Blood 2005;**105**:2640–2643.

42. Blay JY, LeCesne A, Alberti L, Ray-Coquart I. Targeted cancer therapies. Bull Cancer 2005;**92**:E13–E18.

43. McKenzie SB. Advances in understanding the biology and genetics of acute myelocytic leukemia. Clin Lab Sci 2005;**18**:28–37.

44. Goldman J. Monitoring minimal residual disease in BDR-ABL positive chronic myeloid leukemia in the imatinib era. Curr Opin Hematol 2005;**12**:33–39.

45. Melnick A, Licht JD. Deconstructing a disease: RARα, ins fusion partners, and their roles in the pathogenesis of acute promyelocytic leukemia. Blood 1999;**93**:3167–3215.

46. Soignet S, Fleischauer A, Pollyak T, Heller G, Warrel RP Jr. All trans retinoic acid significantly increases 5-year survival in patients with acute promyelocytic leukemia: long term follow-up of the New York study. Cancer Chemother Pharmacol 1997;**40**:S24–S29.

47. Seeler JS, Dehean A. The PML nuclear bodies: actors or extras? Curr Opin Genet Dev 1999;**9**:362–367.

48. Zhong S, Salomoni P, Pandolfi PP. The transcriptional role of PML and the nuclear body. Nat Cell Biol 2000;**2**:E85–E90.

49. Pitha-Rowe I, Petty WJ, Kitareewan S, Dmitrovsky E. Retinoid target genes in acute promyelocytic leukemia. Leukemia 2003;**17**:1723–1730.

50. Melnick A, Licht JD. Deconstructing a disease: RARα, its fusion partners, and their roles in the pathogenesis of acute promyelocytic leukemia. Blood 1999;**99**:3167–3215.

51. Segalla S, Rinaldi L, Kilstrup-Nielsen C, Badaracco G, Minucci S, Pelicci PG, Landsberger N. Retinoic acid receptor alpha fusion to PML affects in transcriptional and chromatin-remodeling properties. Mol Cell Biol 2003;**23**:8795–8808.

52. Dragnev KH, Petty WJ, Dmitrovsky E. Retinoid targets in cancer therapy and chemoprevention. Cancer Biol Ther 2003;**2**(Suppl 1):150–156.

53. Warrell RP Jr, Frankel SR, Miller WH Jr, Scheinberg DA, Itri LM, Hittelman WN, Vyan R, Andreeff M, Tafuri A, Jakubowski A, Gabilove J, Gordon MS, Smitrovsky E. Differentiation therapy of acute promyelocytic leukemia with tretinoin (all-trans-retinoic acid). N Engl Med 1991;**324**:1385–1393.

54. Camacho LH. Clinical application of retinoids in cancer medicine. J Biol Regul Homeost Agents 2003;**17**:98–114.

55. Ohno R, Asou N, Ohnishi K. Treatment of acute promyelocytic leukemia: strategy toward further increase of cure rate. Leukemia 2003;**17**:1454–1463.

56. Parmar S, Tallman MS. Acute promyelocytic leukaemia: a review. Expert Opin Pharmacother 2003;**4**:1379–1392.

57. Tallman MS, Andersen JW, Schiffer CA, Appelbaum FR, Feusner JH, Woods WG, Ogden A, Weinstein H, Shepherd L, Willman C, Bloomfield CD, Rowe JM, Wiernik PH. All-trans-retinoic acid in acute promyelocytic leukemia: long-term outcome and prognostic factor analysis from North American Intergroup protocol. Blood 2002;**100**:4298–4302.

58. Freemantle SJ, Spinella MJ, Dmitrovsky E. Retinoids in cancer therapy and chemoprevention: promise meets resistance. Oncogene 2003;**22**:7305–7315.

59. Chalandon Y, Schwaller J. Targeting mutated protein tyrosine kinases and their signaling pathways in hematologic malignancies. Haematologica 2005;**90**:949–968.

60. Feldman EJ. Farnesyltransferase inhibitors in myelodysplastic syndrome. Curr Hematol Rep 2005;**4**:186–190.

61. Lancet JE, Karp JE. Farnesyltransferase inhibitors in hematologic malignancies: new horizons in therapy. Blood 2003;**102**:3880–3889.

62. Levis M. Recent advances in the development of small-molecule inhibitors for the treatment of acute myeloid leukemia. Curr Opin Hematol 2005;**12**:55–61.

63. Robinson LJ, Xue J, Corey SJ. Src family tyrosine kinases are activated by Flt3 and are involved in the proliferative effects of leukemia-associated Flt3 mutations. Exp Hematol 2005;**33**:469–479.

64. Ono R, Nakajima H, Ozaki K, Kumagai H, Kawashima T, Taki T, Kitamura T, Hayashi Y, Nosaka T. Dimerization of MLL fusion proteins and FLT3 activation synergize to induce multiple-lineage leukemogenesis. J Clin Invest 2005;**115**:919–929.

65. Gilliland DG, Griffin JD. The role of FLT3 in hematopoiesis and leukemia. Blood 2002;**100**:1332–1342.

66. Stone RM, O'Donnell MR, Sekeres MA. Acute myeloid leukemia. Hematology 2004: 98–117.

67. Withers HR, Reid BO, Hussey DH. Response of mouse jejunum to multifraction radiation. Int J Radiat Oncol Biol Phys 1975;**1**:41–52.

68. Lange CS, Gilbert CS. Studies on the cellular basis of radiation lethality. 3. The measurement of stem-cell repopulation probability. Int J Radiat Biol Relat Stud Phys Chem Med 1968;**14**:373–388.

69. Peters R, Layvaz S, Perey L. Apoptotic regulation in primitive hematopoietic precursors. Blood 1998;**92**:2041–2051.

70. Domen J, Gandy KL, Weissman IL. Systemic overexpression of BCL-2 in the hematopoietic system protects transgenic mice from the consequences of lethal irradiation. Blood 1998;**91**:2272–2282.

71. Chaudhary PM, Roninson IB. Expression and activity of P-glycoprotein, a multi-drug efflux pump, in human hematopoietic stem cells. Cell 1991;**66**:85–94.

72. Goodell MA, Brose K, Paradis G, Conner AS, Mulligan RC. Isolation and functional properties of murine hematopoietic stem cells that are replicating in vivo. J Exp Med 1996;**183**:1797–1806.

73. Hirschmann-Jax C, Foster AE, Wulf GG, Nuchtern JG, Jax TW, Gobel U, Goodel MA, Brenner MK. A distinct "side population" of cells with high drug efflux capacity in human tumor cells. Proc Natl Acad Sci U S A 2004;**101**:14228–14233.

74. McCulloch EA. Normal and leukemic hematopoietic stem cells and lineages. In *Stem cells handbook* (Sell S, ed). Totowa, NJ; Humana Press, 2004:119–131.

75. Sutherland HJ, Blair A, Zapf RW. Characterization of a hierarchy in human acute myeloid leukemia progenitor cells. Blood 1996;**87**:4754–4761.

76. Bonnet D, Dick JE. Human acute myeloid leukemia is organized as a hierarchy that originates from a primitive hematopoietic cell. Nature Med 1997;**3**:730–737.

77. Kelly PN, Dakic A, Adams JM, Nutt SL, Strasser A. Tumor growth need not be driven by rare cancer stem cells. Science 2007;**317**:337.

78. Wicha MS, Liu S, Dontu G. Cancer stem cells: an old idea – a paradigm shift. Cancer Res 2006;**66**:1883–1890.

79. Tan BT, Park CY, Ailles LE, Weissman IL. The cancer stem cell hypothesis: a work in progress. Lab Invest 2006;**86**:1203–1207.

80. Trott KR. Tumour stem cells: the biological concept and its application in cancer treatment. Radiother Oncol 1994;**30**:1–5.

81. Denekamp J. Tumour stem cells: facts, interpretation and consequences. Radiother Oncol 1994;**30**:6–10.

82. Pierce GB, Spears WC. Tumors as caricatures of the process of tissue renewal: prospects for therapy by directing differentiation. Cancer Res 1988;**48**:1196–1204.

83. Adams JM, Strasser A. Is tumor growth sustained by rare cancer stem cells or dominant clones? Cancer Res 2008;**68**:4018–4021.

84. Kern SE, Shibata D. The fuzzy math of solid tumor stem cells: a perspective. Cancer Res 2007;**67**:8985–8988.

85. Li C, Heidt DG, Dalerba P, Burant CF, Zhang L, Adsay V, Wicha M, Clarke MF, Simeone DM. Identification of pancreatic cancer stem cells. Cancer Res 2007;**67**:1030–1037.

86. Furth J, Kahn MC. The transmission of leukemia of mice with a single cell. Am J Cancer 1937;**31**:276–282. AN AACR SCIENTIFIC LANDMARK

87. Sell S. Cancer and stem cell signaling: a guide to preventive and therapeutic strategies for cancer stem cells. Stem Cell Rev 2007;**3**:1–6.

88. Hill RP, Perris R. Destemming cancer stem cells. J Natl Cancer Inst 2007;**99**:1435–1440.

89. Ivanova N, Dobrin R, Lu R, Kotenko I, Levorse J, DeCoste C, Schafer X, Lun Y, Lemischka IR. Dissecting self-renewal in stem cells with RNA interference. Nature 2006;**442**:533–538.

90. Sell S (ed). Cancer and stem cell signaling [special issue]. Stem Cell Rev 2007;**3**:1–103.

91. Dreesen O, Brivanlou AH. Signaling pathways in cancer and embryonic stem cells. Stem Cell Rev 2007;**3**:7–17.

92. Farnie G, Clarke RB. Mammary stem cells and breast cancer – role of Notch signaling. Stem Cell Rev 2007;**3**:169–175.

93. Menendez P, Wang L, Bhatia M. Genetic manipulation of human embryonic stem cells: a system to study early human development and potential therapeutic applications. Curr Gene Ther 2005;**5**:375–385.

94. Hay DC, Sutherland L, Clark J, Burdon T. Oct4 knockdown induces similar patterns of endoderm and trophoblast differentiation markers in human and mouse embryonic stem cells. Stem Cells 2004;**22**:225–235.

95. Matin MM, Walsh JR, Gokhale PJ, Draper JS, Bahrami AR, Morton I, Moore HD, Andrews PW. Specific knockdown of Oct4 and beta2-microglobulin expression by RNA interference in human embryonic stem cells and embryonic carcinoma cells. Stem Cells 2004;**22**:659–668.

96. Zaehres H, Lensch ME, Daheron J, Stewart SA, Itskovitz-Eldor J, Daley GQ. High-efficiency RNA interference in human embryonic stem cells. Stem Cells 2005;**23**:299–305.

97. Sprading A, Drummond-Barbarosa D, Kai T. Stem cells find their niche. Nature 2002;**414**:98–104.

98. Wodarz A, Nusse E. Mechanisms of Wnt signaling pathways in development. Annu Rev Cell Dev Biol 1998;**14**:59–88.

99. Tai M-H, Chang C-C, Olson K, Trosko JE. Oct-4 expression in adult human stem cells: evidence in support of the stem cell theory of carcinogenesis. Carcinogenesis 2005;**26**:495–502.

100. Nicholoff BJ, Hendrix MJ, Pollock PM, Trent JM, Miele L, Zin JZ. Notch and NOXA-related pathways in melanoma cells. J Invest Dermatol Symp Proc 2005;**10**:95–104.

101. Callahan R, Egan SE. Notch signaling in mammary development and oncogenesis. J Mammary Gland Biol Neoplasia 2004;**9**:145–163.

102. Kubo M, Nakamura M, Tasaki A, Yamanaka N, Nakashima H, Homura M, Kuroki S, Katano M. Hedgehog signaling pathway is a new therapeutic target for patients with breast cancer. Cancer Res 2004;**64**:6071–6074.

103. Howe LR, Brown AM. Wnt signaling and breast cancer. Cancer Biol Ther 2004;**3**:36–41.

104. Brennan KR, Brown AM. Wnt proteins in mammary development and cancer. J Mammary Gland Biol Neoplasia 2004;**9**:119–131.

105. Milovanovic T, Planutis K, Nguyen A, Marsh JL, Lin F, Hope C, Holcombe RF. Expression of Wnt genes and frizzled 1 and 2 receptors in normal breast epithelium and infiltrating breast carcinoma. Int J Oncol 2004;**25**:1337–1342.

106. Katoh M. Expression and regulation of Wnt-1 in human cancer: up-regulation of Wnt-1 by beta-estradiol in MCF-7 cells. Int J Oncol 2003;**22**:209–212.

107. Saitoh T, Katoh M. Expression and regulation of WNT5A and WNT5B in human cancer: up-regulation of WNT5A by TNFalpha in MKN45 cells and up-regulation of WNT5B by beta-estradiol in MCF-7 cells. Int J Mol Med 2002;**10**:345–349.

108. Katoh M. Regulation of WNT3 and WNT3A mRNAs in human cancer cell lines. Int J Oncol 2002;**20**:373–377.

109. Mishra L, Shetty K, Tang Y, Stuart A, Byers SW. The role of TGF-beta and Wnt signaling in gastrointestinal stem cells and cancer. Oncogene 2005;**24**:5775–5789.

110. Nagae Y, Kameyama K, Yokoyama M, Naito Z, Yamada N, Maeda S, Asano G, Sugisaki Y, Tanaka S. Expression of E-cadherin, catenin and C-erb-2 gene products in invasive ductal-type breast carcinomas. J Nippon Med Sch 2002;**69**:165–171.

111. Jaing WG, Mansel RE. E-cadherin complex and its abnormal sites in human breast cancer. Surg Oncol 2000;**9**:151–171.

112. Lu Z, Ghosh S, Wang Z, Hunter T. Down-regulation of caveolin-1 function by EGF leads to loss of E-cadherin, increased transcriptional activity of β-catenin, and enhanced tumor cell invasion. Cancer Cell 2003;**4**:499–515.

113. Schroeder JA, Adriance MC, McConnell EJ, Thompson MC, Pckaj B, Gendler SJ. ErbB-beta-catenin complexes are associated with human infiltrating ductal breast and murine mammary tumor virus (MMTV)-Wnt-1 and MMTV-c-Neu transgenic carcinomas. J Biol Chem 2002;**277**:22692–22698.

114. Glinsky GV, Berezovska O, Glinskii AB. Microarray analysis identifies a death from cancer signature predicting therapy failure in patients with multiple types of cancer. J Clin Invest 2005;**115**:1503–1521.

115. Glinsky GV. Death-from-cancer signatures and contribution of stem cells to metastatic cancer. Cell Cycle 2005;**4**:1171–1175.

116. Glinsky GV. Genomic models of metastatic cancer: functional analysis of death-from-cancer signature genes reveals aneuploid, anoikis-resistant, metastasis-enabling phenotype with altered cell cycle control and activated Polycomb Group (PcG) protein chromatin silencing pathway. Cell Cycle 2006;**5**:1208–1216.

117. Glinsky GV. Integration of HapMap-based SNP pattern analysis and gene expression profiling reveals common SNP profiles for cancer therapy outcome predictor genes. Cell Cycle 2006;**5**:2613–2625.

118. Liu S, Dontu G, Mantle ID, Patel S, Ahn NS, Jackson KW, Suri P, Wicha MS. Hedgehog signaling and Bmi-1 regulate self-renewal of normal and malignant human mammary stem cells. Cancer Res 2006;**66**:6063–6071.

119. Berezovska OP, Glinskii AB, Yang Z, Li XM, Hoffman RM, Glinsky GV. Essential role for activation of the Polycomb group (PcG) protein chromatin silencing pathway in metastatic prostate cancer. Cell Cycle 2006;**5**:1886–1901.

120. Zhou J, Zhang H, Gu P, Bai J, Margolick JB, Zhang Y. NF-kappaB pathway inhibitors preferentially inhibit breast cancer stem-like cells. Breast Cancer Res Treat 2008;**111**(3):419–427.

121. Rye PD, Stigbrand T. Interfering with cancer: a brief outline of advances in RNA interference in oncology. Tumor Biol 2004;**25**:329–336.

122. Sorensen DR, Leirdal M, Sioud M. Gene silencing by systemic delivery of synthetic siRNAs in adult mice. J Mol Biol 2003;**327**:761–776.

123. Pirollo KF, Chang EH. Targeted delivery of small interfering RNA: approaching effective cancer therapies. Cancer Res 2008;**68**:1247–1250.

124. Pirollo KF, Rait A, Zhou Q, Hwang SH, Dagata JA, Zon G, Hogrefe RI, Palchik G, Chang EH. Materializing the potential of small interfering RNA via tumor-targeting nanodelivery system. Cancer Res 2007;**67**:2938–2943.

125. Zamecnik PC, Stephenson ML. Inhibition of Rous sarcoma virus replication and cell transformation by a specific ologodeoxynucleotide. Proc Natl Acad Sci U S A 1978;**75**:280–284.

126. Harrington KJ, Nutting CM, Pandha HS. Gene therapy for head and neck cancer. Cancer Metastasis Rev 2005;**24**:147–164.

127. Skorski T, Neiborowska-Skorska M, Barletta C, Malaguamera L, Szczylik C, Chen S-T, Lang B, Calabretta B. Highly efficient elimination of Philadelphia1 leukemic cells by exposure to bcr/abl antisense oligodeoxynucleotides combined with mafosfamide. J Clin Invest 1993;**92**:194–202.

128. Liebermann TA, Zerbini LF. Targeting transcription factors for cancer gene therapy. Curr Gene Ther 2006;**6**:17–33.

129. Hanazono Y, Asano T, Ueda Y, Ozawa K. Genetic manipulation of primate embryonic and hematopoietic stem cells with simian lentivirus vectors. Trends Cardiovasc Med 2003;**13**:371–378.

130. Ikawa M, Tanaka N, Kao WW, Verma IM. Generation of transgenic mice using lentiviral vectors: a novel preclinical assessment of lentiviral vectors for gene therapy. Mol Ther 2003;**8**:666–673.

131. Al-Haji M, Wicha MS, Benito-Hernandez A, Morrison SJ, Clarke MR. Prospective identification of tumorigenic breast cancer cells. Proc Natl Acad Sci U S A 2003;**100**:3983–3988.

132. Shipitsin M, Campbell LL, Argani P, Weremowicz S, Bloushtain-Qimron N, Yao J, Nikolskaya T, Serebryiskaya T, Beroukhim R, Hu M, Halushka MK, Sukumar S, Parker LM, Anderson KS, Harris LN, Garber JE, Richardson AL, Schnitt SJ, Nikolsky Y, Gelman RS, Polyak K. Molecular definition of breast tumor heterogeneity. Cancer Cell 2007;**11**:259–273.

133. Korsching E, Jeffrey SS, Meinerz W, Decker T, Boecker W, Buerger H. Basal carcinoma of the breast revisited: an old entity with new interpretations. J Clin Pathol 2008;**61**:553–560.

134. Roskoski R Jr. Structure and regulation of Kit protein-tyrosine kinase – the stem cell factor receptor. Biochem Biophys Res Commun 2005;**338**:1307–1315.

135. Lennartsson J, Jelacic T, Linnekin D, Shivakrupa R. Normal and oncogenic forms of the receptor tyrosine kinase Kit. Stem Cells 2005;**23**:16–43.

136. van de Bijver M. Gene-expression profiling and the future of adjuvant therapy. Oncologist 2005;**10**(Suppl 2):30–34.

137. Toren A, Bielora B, Jacob-Hirsch J, Fisher T, Kreiser D, Moran O, Zeligson S, Givol D, Yitzhaky-Eldor J, Kventsel I, Rosenthal E, Amarigilio N, Rechavi R. CD133-positive hematopoietic stem cell "stemness" genes contain many genes mutated or abnormally expressed in leukemia. Stem Cells 2005;**23**:1142–1153.

138. Dar A, Goichberg P, Shinder V, Kalinkovich A, Kollet O, Netzer N, Margalit R, Zsak M, Nagler A, Hardan I, Resnick I, Rot A, Lapidot T. Chemokine receptor CXCR4-dependent internalization and resecretion of functional chemokine SDF-1 by bone marrow endothelial and stromal cells. Nature Immunol 2005;**6**:1038–1046.

139. Weissman IL. Stem cells: units of development, units of regeneration, and units in evolution. Cell 2000;**100**:157–168.

140. Kiel MJ, Yilmaz OH, Iwashita T, Yilmaz OS, Terhorst C, Morrison SJ. SLAM family receptors distinguish hematopoietic stem and progenitor cells and reveal endothelial niches for stem cells. Cell 2005;**121**:1109–1121.

141. Sell S. Oncodevelopmental antigens: a review. Cancer Biol Rev 1980;**1**:251–352.

142. Nomura K, Nagano K, Itagaki C, Raoka M, Okamura N, Yamauchi Y, Sugano S, Takahashi N, Izumi T, Isobe T. Cell surface labeling and mass spectrometry reveal diversity of cell surface markers and signaling molecules expressed in undifferentiated mouse embryonic stem cells. Mol Cell Proteomics 2005;**4**:1968–1976.

143. Deicher R, Hori WH. Differentiating factors between erythropoiesis-stimulating agents: a guide to selection for anemia of chronic kidney disease. Drug 2004;**54**:499–509.

144. Nowell PC. The clonal evolution of tumor cell populations. Science 1976;**194**:23–28.

6 Targeting acute myelogenous leukemia stem cells

Monica L. Guzman
University of Rochester School of Medicine and Dentistry

Gerrit J. Schuurhuis
VU University Medical Center

Craig T. Jordan
University of Rochester Medical Center

Stem cells are an important part of the physiology of different tissues, including the hematopoietic system. Hematopoietic stem cells (HSC) serve as an essential component of the blood-forming system and generate all cell types found in mature blood. Not surprisingly, given the importance of HSCs, their phenotypes and properties have been studied extensively in both human and murine systems. Interestingly, by analogy to HSCs, a large body of evidence has shown that different types of leukemia can arise from a malignant counterpart to the HSC, known as a leukemia stem cell (LSC) or leukemia-initiating cell (L-IC).[1–4] Recent studies suggest that LSCs are not effectively eliminated by commonly utilized therapeutic regimens and thereby represent a likely source of leukemia relapse. Consequently, there is an increasing interest in understanding LSC biology to identify a means of ablating malignant stem cells. Due to the extensive understanding of the normal hematopoietic system, and the availability of tools and techniques to isolate and assay this population of cells, substantial progress has been made both in understanding the properties of LSCs and in the development of promising selective anti-LSC therapies.

Table 6–1. Leukemia stem cells for chronic myelogenous leukemias and B and T cell acute lymphoblastic leukemias

Type of leukemia	Phenotypically described LSC	References
Acute myelogenous leukemia[a]	CD34+ CD38− CD90− CD117− CD123+ CD96+ CLL-1+	(1,3,16−19)
Blast crisis chronic myelogenous leukemia	CD34+ CD38+	(29)
T-cell acute lymphoblastic leukemia	CD34+ CD4−; CD34+ CD7−	(87)
B-cell acute lymphoblastic leukemia	CD34+ CD38−/lowCD19+; CD34+ CD38− CD33− CD19−; CD34+ CD38−; CD34+ CD10− CD19−; CD133+ CD38− CD19−	(13,88−91)

[a] Not all cases obey the indicated immunophenotype.

IDENTIFICATION OF LSCs IN HUMAN LEUKEMIAS

In the 1960s, Fialkow and colleagues provided supporting evidence for the stem cell origin of human leukemia by studying X-linked glucose-6-phosphate dehydrogenase (G-6-PD) in erythrocytes and granulocytes from patients diagnosed with chronic myelogenous leukemia (CML).[5] It was found in female patients heterozygous for two variant types of G-6-PD that only one of the two variants was present in both erythrocytes and granulocytes. This observation strongly suggested that a common stem cell gave rise to these two cell populations. Later, with the identification of the Philadelphia chromosome (Ph), t(9;22), in CML, further evidence for the involvement of stem cells in the disease was obtained.[6] Indeed, multilineage involvement has been described for Ph+ CML and acute lymphoblastic leukemia (ALL) patients.[6,7]

The first demonstration that a small subpopulation of cells within a heterogeneous tumor can recapitulate leukemic disease was reported by Lapidot and colleagues in 1994.[1] Those studies employed an immunodeficient mouse xenotransplant assay for tumor initiation, which is now the gold standard for the validation of cancer stem cell function in vivo. Subsequent studies by Bonnet and Dick[2] provided a broad analysis of multiple subtypes of acute myelogenous leukemia (AML) and included detailed stem cell analysis. It was observed that a rare population of cells (~0.1% to 1%) is present that displays similar phenotypic markers to normal HSCs (CD34+/CD38−). This CD34+/CD38− subpopulation was shown to initiate AML and recapitulate disease when transplanted into nonobese diabetic/severe combined immunodeficiency disease (NOD/SCID) mice. Subsequent studies identified LSCs for chronic myelogenous leukemias as well as B and T cell acute lymphoblastic leukemias (Table 6–1). The identification of LSC populations in different types of leukemia, and the definition of their phenotypic markers, has served as a foundation for the study of their role in pathogenesis as well as response to therapies.

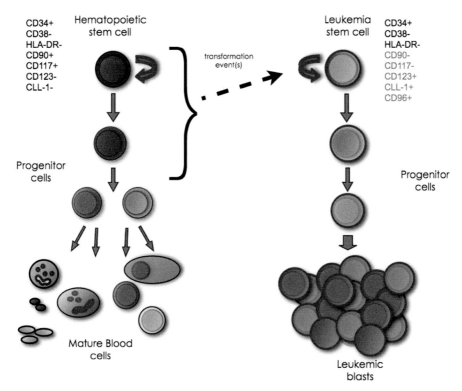

Figure 6–1: Leukemia stem cells (LSCs) are thought to arise from a normal hematopoietic stem cell or progenitor cell that undergoes a malignant transformation event. LSCs then have the ability to give rise to leukemia progenitor and blast populations. *See color plates.*

CELL OF ORIGIN

Models for LSC origin suggest that such cells may arise from normal hematopoietic stem or progenitor cells that undergo a series of transformation events that perturb self-renewal and differentiation pathways (Figure 6–1). As a result, these transformed cells acquire the ability to give rise to an aberrant population of immature cells known as *leukemic blasts*.

Different theories for the cellular origin of leukemia have arisen from detailed studies of disease-specific translocations and the prospective isolation and identification of phenotypic markers for LSCs: (1) in chronic-phase CML patients, the characteristic BCR-ABL translocation has been found in different hematopoietic lineages, supporting the idea that the disease originates at the HSC; (2) in AML, a very heterogeneous disease both phenotypically and clinically, most subtypes have been shown to contain LSCs with similar phenotypes that are to some extent similar to the normal HSCs (CD34+, CD38−) – this similarity was taken to suggest that LSCs may originate from a transformed HSC[1,8]; (3) in contrast, specific lineage restrictions observed for some AML patients, such as monocytic/granulocytic leukemia, have suggested the possibility of malignant transformation in a committed progenitor cell as the origin for disease.[9]

More recent studies have strongly suggested the possibility that the transformation event can occur sequentially in different developmental stages, where an initial mutation originating in a stem cell creates a preleukemic LSC. Subsequent mutations at either the stem or progenitor cell level transform the preleukemia stem cell into an LSC that can give rise to frank leukemia,[10] consistent with the classic two-hit model of oncogenesis.[11] For example, in phenotypically primitive cells (CD34+CD38−CD90+) isolated from patients with AML-ETO translocation (t(8;21)) AML in remission, it was shown that even though AML-ETO transcripts were detected, the cells formed normal multilineage colonies with no disease phenotype. In contrast, more mature cells (CD34+CD38+) give rise to leukemic blast cells.[12] These observations suggest that the initial transformation (AML-ETO) occurred in a stem cell, but that subsequent mutations likely occurred at later stages of development, thereby giving rise to a fully transformed leukemic stem/progenitor. Further support for this model arises from a study of identical twins.[13] One twin presented with B-cell ALL bearing the translocated leukemia (TEL)-AML chromosomal translocation, while the healthy twin carried a very rare population of apparently premalignant cells (0.002%) with the TEL-AML translocation.[13] These studies indicate that disease pathogenesis may involve a preleukemia stem cell stage. Leukemia mouse models have additionally shown that the nature of the transformation event also plays a role in the generation of preleukemia and leukemia stem cells in different hematopoietic stem and progenitor pools.[9,14,15]

It is likely that preleukemia stem cells possess different biological characteristics that distinguish them from LSC and leukemia blast populations. Thus, to improve therapeutic targeting and assessment of minimal residual disease, a better understanding of the characteristics of these populations and their role in leukemogenesis is essential.

CHARACTERISTICS

Since the initial functional definition of AML LSCs in 1994, there has been a continuous effort to better characterize these cells at the phenotypic and molecular levels. LSCs share some immunophenotypic markers with normal HSCs (CD34+CD38−). However, certain differences have also been described. Unlike HSCs, LSCs generally do not express CD90 and CD117, but are usually positive for CD123, CD96, and CLL-1.[3,16–19] Together, these phenotypic differences allow for the detection, isolation, and characterization of LSCs. In addition to an aberrant phenotype, LSCs may have altered self-renewal and survival characteristics when compared to their normal HSC counterparts. Using lentiviral marking of LSCs and serial transplantation in immunodeficient mice, LSC populations were shown to be heterogeneous, consisting of cells with different self-renewal capacities. In some cases, potentially higher self-renewal capacity than found in normal HSCs was discovered.[20]

The cellular and molecular analysis of LSCs has been greatly facilitated by the ability to perform prospective isolation and functional analysis. Unlike actively cycling bulk tumor cells, LSCs often exist in a quiescent state, a characteristic that is shared with their normal counterparts.[21,22] This state of quiescence suggests that LSCs will likely be more refractory to commonly used therapies. Indeed, the relative resistance of LSCs to chemotherapy has been demonstrated experimentally for agents, such as cytarabine and daunorubicin, that are designed to interfere with DNA replication and/or repair.[23–25]

Importantly, a number of studies have identified constitutively activated survival pathways in LSCs that are not active in normal HSCs.[26,27] The activation of survival pathways in LSCs has been exploited as a means to preferentially target LSCs and not HSCs. One such survival mechanism is the NF-κB pathway, which is found constitutively activated in leukemia cells and, interestingly, also in the quiescent LSC populations.[26] In addition, the PI3 kinase signaling pathway has been found to be constitutively active in primary AML samples, and inhibition of this pathway also impairs LSCs.[27,28]

As mentioned previously, in addition to altered survival, LSCs can also demonstrate aberrant activation of self-renewal capacity. An example of this state has been shown for granulocyte-macrophage progenitors from patients with blast crisis CML, where activated beta-catenin contributes to their increased self-renewal ability.[29] This capacity can be reduced by overexpression of Axin to inhibit beta-catenin.

LSCs AND THERAPY

Increasing evidence suggests that most current therapeutic regimens fail to eradicate LSCs.[30–34] Thus some consideration of LSC biology is essential for developing antileukemia therapies that will more effectively target malignant stem cells. For example, agents, such as cytarabine, that are very effective in targeting actively cycling cells fail to ablate the quiescent LSCs.[23] Similarly, in CML, while the use of imatinib (BCR-ABL kinase-specific inhibitor) effectively eradicates bulk disease, the drug is cytostatic only for primitive cells and thereby unable to eliminate LSCs.[33,34] Consequently, there has been an increased effort to identify therapies that can more effectively target primitive leukemic cells, exploiting a diverse set of features that discriminate LSCs from HSCs.

Phenotypic markers

CD123, the IL-3 receptor alpha, is expressed in LSCs from AML patients but is not found on normal HSCs. A potential approach to exploit this difference is the use of fusion constructs of the catalytic and translocation domains of diphtheria toxin (DT_{388}) to human IL-3. The fusion has been found to selectively kill AML cells in vitro[35–38] and in a mouse xenograft model system.[38] Furthermore, in phase

I clinical trials, AML cytoreductions were observed in 17.5% of patients. However, since the sensitivity to the DT_{388}IL-3 varies with the amount of expression of the high-affinity IL-3R heterodimers (alpha and beta),[39] and the expression of the IL-3R beta chain (CD131) has not been well described in primitive cells, the net effect of DT_{388}IL-3 on LSCs requires further investigation.

CD33 expression in LSCs from AML patients has been a matter of debate. There are reports that describe the expression of CD33 in primitive AML populations.[40,41] Another report suggested that CD34+ progenitors from AML patients do not express CD33.[42] CD33 was, however, also found expressed in normal CD34+CD38-Lineage− (lineage: CD3, CD14, CD16, CD19, CD20, CD56),[40] providing a possible explanation for cytopenias observed with anti-CD33 therapy. Gemtuzumab ozogamicin (Mylotarg) is an anti-CD33 monoclonal antibody linked to the toxic agent calicheamicin and is currently approved by the U.S. Food and Drug Administration for treatment of elderly patients with relapsed AML who are not candidates for standard chemotherapy.[43] The clinical use of Mylotarg provides an avenue of investigation for the response in LSC populations in patients who have received treatment.

CD44 expression has been found in CD34+CD38− cells from AML patients and was recently shown to be an important regulator of LSC fate, as demonstrated by treatment with an anti-CD44 activating antibody (H90), which reduced engraftment of primary AML cells in immunodeficient NOD/SCID mice.[44] In addition, studies of CD44 in a mouse model of CML have shown that LSCs are dependent on CD44 for homing and engraftment more than normal HSCs, emphasizing the importance of CD44 for LSC targeting.[45]

CLL-1 and CD96 have both been recently demonstrated to be differentially expressed in LSCs from AML patients.[16,17] Ongoing studies will determine whether these epitopes represent useful targets for therapy.

CXCR4, also know as fusin, is a chemokine (CXC) receptor for SDF-1 (CXCL12). CXCR4/SDF-1 interactions have an important role in the homing of HSCs to the bone marrow.[46] In AML, it has been reported that low expression of CXCR4 correlated with longer relapse-free survival.[47] The use of CXCR4 neutralizing antibodies or AMD3100 (CXCR4 antagonist) significantly decreased the survival of AML cells in vitro. Moreover, blocking of CXCR4 also resulted in decreased engraftment of primary AML cells.[48] Since the interactions of LSCs with the bone marrow microenvironment are likely conferring additional survival signals, allowing them to escape from chemotherapeutic insults, the use of approaches that disrupt these interactions are of considerable interest.

Molecular targets

An alternative strategy for ablation of LSCs is to target pathways involved in survival and pathogenesis that are unique to primitive leukemia cells. However, due to the substantial genetic and phenotypic heterogeneity found among leukemia patients, the identification of common pathways may be difficult. Nonetheless, several pathways have emerged in recent years that may be relevant to LSC biology and represent potential targets for therapy.

FLT3 (FMS-like tyrosine kinase) is a tyrosine kinase receptor involved in proliferation, survival, and differentiation of normal hematopoietic progenitor cells.[49] Activating mutations, such as internal tandem duplications (ITD) or point mutations in the kinase domain, have been found in AML and are often correlated with poor survival rates.[50-53] In addition, it has been shown that FLT3-ITD mutations are observed in the primitive CD34+CD38– cells from AML patients harboring FLT3-ITD mutations.[54] While the biological relevance of FLT3 activation in LSCs has not yet been determined, growing interest in the use of FLT3 inhibitors for the treatment of AML has provided an opportunity to address this issue in malignant stem and progenitor cell populations.[54,55]

PI3K (phosphatidyl-inositol 3 kinase) is a lipid kinase regulated in response to multiple hematopoietic cytokines and chemokines. It serves as a broad regulator of cell proliferation and cell growth. The effects of PI3K on cell survival are mediated through its downstream effector, the serine-threonine kinase protein kinase B (PKB)/Akt.[56] It has been shown that Akt is constitutively active in most primary AML specimens. Furthermore, mTOR (mammalian target of rapamycin), the downstream effector of Akt, was shown to be necessary for survival of LSCs following genotoxic stress.[27,28] Treatment with the PI3K inhibitor LY294002 resulted in growth inhibition and apoptosis of AML cells with minimal effect on normal CD34+ hematopoietic progenitors.[27] The importance of this pathway for the survival of LSCs was underscored by the decreased ability of cells to engraft NOD/SCID mice after treatment with LY294002. Currently, rapamycin derivatives, such as everolimus, temsirolimus, and AP23573, as well as emerging PI3K inhibitors are being evaluated in preclinical and clinical trials for different types of malignancies.[57] Obtaining safety and pharmacodynamic data on these types of drugs will indicate whether inhibition of mTOR can sensitize LSCs in vivo to the genotoxic stress induced by most common therapeutic agents, as suggested by in vitro studies.

NF-κB is a transcription factor involved in cell growth and survival and has been found to be constitutively activated in many types of cancers, including leukemia.[58-63] Surprisingly, it is also active in the quiescent LSC populations of primary AML samples.[26] In contrast, normal CD34+ cells do not activate NF-κB at steady state. Thus strategies to inhibit NF-κB represent an interesting approach to more durable AML therapy. To this end, different approaches have been explored. Proteasome inhibitors block the degradation of IκBα, the endogenous inhibitor of NF-κB, thereby resulting in the loss of NF-κB activity. Initial studies in primary AML cells demonstrated that treatment with the proteasome inhibitor carbobenzoxyl-L-leucyl-L-leucyl-L-leucinal (MG-132) induced robust apoptosis in less than 24 hours.[26] Molecular genetic studies using a dominant-negative allele for IκBα demonstrated that inhibition of NF-κB alone is not sufficient to induce robust cell death of primary AML cells; however, it likely contributes to the apoptotic process. Further analyses of MG-132 in combination with the anthracycline idarubicin showed improved targeting of LSC and suggested that activation of p53-regulated genes is also involved in LSC apoptosis.[23]

Another approach to targeting the NF-κB pathway is use of the sesquiterpene lactone parthenolide (PTL). This compound is derived from a medicinal plant known as feverfew. PTL has been shown to specifically target LSCs from AML and blast crisis CML patients while sparing normal HSCs.[64] While in vitro studies of PTL showed promising anti-LSC activity, in vivo studies with this drug have been limited by the poor pharmacological properties of the compound.[30] To overcome this limitation, structural analogs of PTL have been created to generate a more useful clinical agent. From these efforts, a dimethyl-amino analog with over a 1,000-fold greater solubility than PTL was identified. The compound, known as DMAPT or LC-1, has 70% oral bioavailability in rodent and canine models. DMAPT activity against LSC was also documented, as was the activity to inhibit NF-κB in vivo.[30] Interestingly, DMAPT also showed the ability to induce differentiation of canine leukemia cells in vivo, although these observations need to be further corroborated in a larger and more detailed study. Future clinical studies will assess whether DMAPT can target LSCs in human leukemia patients.

Other approaches

Upregulation of antiapoptotic proteins, such as Bcl-2 and Bcl-xl, is commonly found in cancers.[65] Therefore inhibitory molecules for Bcl-2 and Bcl-xl have been developed. One such inhibitor is ABT-737, which is a small molecule that mimics BH3-domain molecules, such as Bax and Bak, that naturally bind to Bcl-2 and Bcl-xl, resulting in cell death.[66,67] Recently, it was shown that ABT-737 can kill primary leukemia cells with little effect on normal hematopoietic cells. The study also showed activity of ABT-737 against phenotypically defined AML stem cells, thereby indicating potential utility of the drug for LSC-targeted therapies.[67]

TDZD-8 (4-benzyl-2-methyl-1,2,4-thiadiazolidine-3,5-dione) was originally described as a GSK3-beta inhibitor.[68] However, a recent report revealed that TDZD-8 demonstrates the unique ability to induce cell death of primary LSCs from AML, ALL, and blast crisis CML patients within 2 hours after drug exposure. In addition, no apparent toxicity was observed in normal HSCs.[69] The activity of TDZD-8 appears to be independent of GSK3-beta since other GSK3-beta inhibitors do not demonstrate such properties. TDZD-8 was found to inhibit several kinases involved in growth and survival functions of malignant cells such as PKC family members, Akt, FLT3, and Aurora B. Although the primary target of TDZD-8 is not yet clear, the specificity and kinetics of the drug make this agent of unique interest for LSC targeting.

In silico screening of publicly available gene expression profiles in the Gene Expression Omnibus (GEO) has been used to identify compounds that can differentially target LSCs.[70] Specifically, the gene expression pattern from PTL treatment of primary CD34+ AML samples was used as a probe for identifying similar patterns in other biological states. From this search, two new compounds with anti-LSC activity, both of which are chemically distinct from PTL, were identified:

celastrol and 4-hydroxynonenal. While these compounds require pharmacologic optimization for clinical use, the process nonetheless represents a potentially powerful means to screen for novel antileukemia drugs in a systematic manner.

Other possible targets

Notch, Hedgehog, and Wnt signaling pathways are important regulators of stem cell self-renewal and differentiation. The aberrant activation of these pathways has been described in studies of numerous malignancies.[71–75] Notch and Wnt signaling pathways have also been shown to play a role in the regulation of LSCs obtained from blast crisis CML, which demonstrate the nuclear accumulation of β-catenin, a downstream effector of the Wnt signaling cascade.[29] Moreover, Jagged-2, a ligand for Notch, has been found to be overexpressed in LSCs from AML patients.[76] Given the potential relevance to cancer, there is an increasing interest from pharmaceutical companies to develop specific inhibitors against molecules involved in these pathways. These agents are currently at early stages of development, in either preclinical studies or phase I trials. While it is unclear whether such drugs will effectively discriminate between normal and malignant stem cells, their study represents an important line of investigation.

LSCs AND CLINICAL IMPLICATIONS

As outlined previously, therapies aimed at eradication of LSCs ideally should target a characteristic of the LSC that is absent in the normal HSC.

Irrespective of the time point during treatment/disease (i.e., diagnosis, after a therapy course, or at follow-up after the last therapy course), normal HSCs coexist with LSCs: at diagnosis, normal stem cells may make up only a fraction of the total stem cell compartment, while in remission, in many cases, the vast majority of stem cells will be normal.

Diagnostic approaches, be they aiming at therapeutic targeting of LSCs or discovering new targets, should therefore take into account the presence of these two populations, which may widely differ both in response to current therapies or by the presence of newly defined therapeutic targets. These approaches demand the ability to discriminate both types of stem cells while in viable condition. As mentioned previously, known differences in cell surface marker expression allow identification of LSC. In AML, CD123,[19] CD44,[44] CD96,[17] and CLL-1[16] have been described as stem cell markers staining the majority of the CD34+CD38− stem cell compartment, which, as such, has a frequency of 0.1% to 1% in total blast cell populations. Study of expression on normal bone marrow stem cells of healthy volunteers (or patients with non–bone marrow diseases) revealed the absence of CLL-1[16] and CD44,[44] low expression of CD96[17] and CD123,[19,77] and prominent expression of CD33.[40,77]

An increasing number of reports deal with effects of leukemia blast cells on coexisting normal cells in leukemia bone marrow. Although stem cells were not part of these studies, the results suggest that nonmalignant cells may not be as normal as regularly used normal bone marrow controls: these nonmalignant cells may acquire immunologic and apoptosis characteristics that resemble those of their malignant counterparts.[78] If such would occur for a target identified as relevant for therapy by comparison with normal bone marrow stem cells, it may abrogate the desired specificity for the AML stem cell. Although more difficult for diagnostic, therapeutic, and discovery purposes, it thus seems far more appropriate to use the nonmalignant stem cells present in the leukemia bone marrow as the reference normal stem cell pool.

Apart from the above-mentioned markers, a number of lineage markers, that is, B-cell, T-cell, NK-cell, and myeloid cell markers, have been found to characterize the CD34+CD38− AML stem cell compartment at diagnosis. These markers are absent on normal bone marrow stem cells as well as on nonmalignant stem cells in the AML bone marrow.[77] Moreover, and in contrast to CD123 and CLL-1, the markers are virtually absent on the nonmalignant CD34+CD38+ progenitor compartment, too, which was recognized years ago and provides the basis for immunophenotypic minimal residual disease (MRD) detection (see subsequent discussion).

Additional differences between normal and leukemic stem cell properties may include cell size and granularity, with AML stem cells being slightly larger and more granular than more mature AML cell types (Terwijn, M, unpublished observations). Such observations are not restricted to AML: in CML, as well, based on differences in marker expression and scatter properties, and together with fluorescent in situ hybridization (FISH) analysis, the CD34+CD38− stem cell compartment reveals the presence of coexisting normal and leukemic stem cells (Janssen, JJWM, unpublished observations).

Remission bone marrow evaluation often reveals the presence of malignant blasts (not discriminating between stem cells and more differentiated cells); these MRD cells can be quantified using either polymerase chain reaction or immunophenotypic techniques.[79–82] In the latter case, immunophenotypic aberrancies, such as cross-lineage antigen expression, antigen overexpression, or asynchronous antigen expression, are defined at diagnosis and used for follow-up to quantify MRD and in prognosis. Although MRD cell frequency is one of the best overall predictors of relapse,[79–82] similar to other prognostic markers, this parameter is not always useful for individual patients. Sequential bone marrow biopsies showing increased MRD frequency can be used for prognosis, but this is a burden for the patient and time consuming for the clinician.

Since the size of the stem cell compartment at diagnosis has been shown to be predictive for the amount of MRD after treatment as well as for clinical outcome,[83] it may be argued that stem cell frequency assessment could contribute to the prognostic impact in cases with discrepancies between MRD cell frequencies and clinical outcome. Furthermore, therapeutic success may be related not only to the frequency of AML stem cells, but also to the balance of AML and normal stem cells competing for the stem cell niche.

Indeed, by using marker expression and scatter parameters, as discussed earlier, AML and normal stem cells can be discriminated under remission conditions and in preliminary evaluations, showing prognostic impact.[16,77] Care must be taken with those markers that may be upregulated on normal HSCs present in regenerating bone marrow of the patient.

With clinically relevant MRD blast cell frequencies of 0.01% to 0.1%,[79–82] and with AML-SCs making up 0.1% to 1% of AML blast cells, in the majority of patients, AML-SCs will present with frequencies in the range of 1:100,000 to 1:10 million. Although such frequencies, in general, do not allow functional or molecular characterization of cells isolated by flow sorting, in vitro testing of stem cell therapies is possible using analytical flow cytometry of the bulk of bone marrow cells, combined with appropriate markers for MRD cells and stem cells as readouts for efficacy.

Although the commonly used definition of the AML stem cell compartment is by immunophenotype, evidence for other definitions has also been reported. For example, a so-called side population (SP) of cells, defined by high extrusion of the DNA binding drug Hoechst 33342, has been described in normal bone marrow as containing HSC activity.[84] Similarly, in human AML, the SP population contains leukemia-initiating cells that read out in the NOD/SCID xenograft model.[85,86] While use of the SP phenotype may augment conventional cell surface markers, its significance at clinical follow-up and its role in predicting outcome are as yet unknown.

SUMMARY

The elucidation of specific markers and biological characteristics of human LSCs has provided exciting opportunities for the development of new forms of therapeutic intervention. LSCs can now be physically isolated and subjected to most forms of experimental analysis, thereby allowing detailed studies of relative drug sensitivity and response. Furthermore, with the availability of advanced xenograft models, it is also possible to perform in vivo analyses of LSCs in which their biological properties can be further defined. As a consequence of these capabilities, a variety of new therapeutic strategies for targeting LSCs have emerged, some of which are undergoing clinical testing. These approaches include the use of several different small molecules, antibody-based drugs, and cytokine–toxin fusion proteins. Going forward, the evaluation of such therapies will rely on developing sensitive assays for the detection and quantification of LSCs in patients undergoing treatment with candidate targeted regimens. While assessing primitive leukemia cells in vivo represents a significant technical and logistical challenge, it appears that all necessary tools are in place to effectively monitor LSC burden in patients, and to thereby determine the relevance of targeting LSCs with respect to clinical outcome.

REFERENCES

1. Lapidot T, Sirard C, Vormoor J, et al. A cell initiating human acute myeloid leukaemia after transplantation into SCID mice. Nature. 1994;367:645–648.

2. Bonnet D, Dick JE. Human acute myeloid leukemia is organized as a hierarchy that originates from a primitive hematopoietic cell. Nat Med. 1997;3:730–737.

3. Blair A, Hogge DE, Ailles LE, et al. Lack of expression of Thy-1 (CD90) on acute myeloid leukemia cells with long-term proliferative ability in vitro and in vivo. Blood. 1997;89:3104–3112.

4. Sutherland HJ, Blair A, Zapf RW. Characterization of a hierarchy in human acute myeloid leukemia progenitor cells. Blood. 1996;87:4754–4761.

5. Fialkow PJ, Jacobson RJ, Papayannopoulou T. Chronic myelocytic leukemia: clonal origin in a stem cell common to the granulocyte, erythrocyte, platelet and monocyte/macrophage. Am J Med. 1977;63:125–130.

6. Dube ID, Gupta CM, Kalousek DK, et al. Cytogenetic studies of early myeloid progenitor compartments in Ph1-positive chronic myeloid leukaemia (CML). I. Persistence of Ph1-negative committed progenitors that are suppressed from differentiating in vivo. Br J Haematol. 1984;56:633–644.

7. Schenk TM, Keyhani A, Bottcher S, et al. Multilineage involvement of Philadelphia chromosome positive acute lymphoblastic leukemia. Leukemia. 1998;12:666–674.

8. Jordan CT. Unique molecular and cellular features of acute myelogenous leukemia stem cells. Leukemia. 2002;16:559–562.

9. Krivtsov AV, Twomey D, Feng Z, et al. Transformation from committed progenitor to leukaemia stem cell initiated by MLL-AF9. Nature. 2006;442:818–822.

10. Yuan Y, Zhou L, Miyamoto T, et al. AML1-ETO expression is directly involved in the development of acute myeloid leukemia in the presence of additional mutations. Proc Natl Acad Sci USA. 2001;98:10398–10403.

11. Moolgavkar SH, Knudson AG Jr. Mutation and cancer: a model for human carcinogenesis. J Natl Cancer Inst. 1981;66:1037–1052.

12. Miyamoto T, Weissman IL, Akashi K. AML1/ETO-expressing nonleukemic stem cells in acute myelogenous leukemia with 8;21 chromosomal translocation. Proc Natl Acad Sci USA. 2000;97:7521–7526.

13. Hong D, Gupta R, Ancliff P, et al. Initiating and cancer-propagating cells in TEL-AML1-associated childhood leukemia. Science. 2008;319:336–339.

14. Huntly BJ, Shigematsu H, Deguchi K, et al. MOZ-TIF2, but not BCR-ABL, confers properties of leukemic stem cells to committed murine hematopoietic progenitors. Cancer Cell. 2004;6:587–596.

15. Neering SJ, Bushnell T, Sozer S, et al. Leukemia stem cells in a genetically defined murine model of blast-crisis CML. Blood. 2007;110:2578–2585.

16. van Rhenen A, van Dongen GA, Kelder A, et al. The novel AML stem cell associated antigen CLL-1 aids in discrimination between normal and leukemic stem cells. Blood. 2007;110:2659–2666.

17. Hosen N, Park CY, Tatsumi N, et al. CD96 is a leukemic stem cell-specific marker in human acute myeloid leukemia. Proc Natl Acad Sci USA. 2007;104:11008–11013.

18. Blair A, Sutherland HJ. Primitive acute myeloid leukemia cells with long-term proliferative ability in vitro and in vivo lack surface expression of c-kit (CD117). Exp Hematol. 2000;28:660–671.

19. Jordan CT, Upchurch D, Szilvassy SJ, et al. The interleukin 3 receptor alpha chain is a unique marker for human acute myelogenous leukemia stem cells. Leukemia. 2000;14:1777–1784.

20. Hope KJ, Jin L, Dick JE. Acute myeloid leukemia originates from a hierarchy of leukemic stem cell classes that differ in self-renewal capacity. Nat Immunol. 2004;5:738–743.

21. Holyoake TL, Jiang X, Jorgensen HG, et al. Primitive quiescent leukemic cells from patients with chronic myeloid leukemia spontaneously initiate factor-independent growth in vitro in association with up-regulation of expression of interleukin-3. Blood. 2001;97:720–728.

22. Guan Y, Gerhard B, Hogge DE. Detection, isolation, and stimulation of quiescent primitive leukemic progenitor cells from patients with acute myeloid leukemia (AML). Blood. 2003;101:3142–3149.

23. Guzman ML, Swiderski CF, Howard DS, et al. Preferential induction of apoptosis for primary human leukemic stem cells. Proc Natl Acad Sci USA. 2002;99:16220–16225.
24. Ishikawa F, Yoshida S, Saito Y, et al. Chemotherapy-resistant human AML stem cells home to and engraft within the bone-marrow endosteal region. Nat Biotechnol. 2007;25:1315–1321.
25. Costello RT, Mallet F, Gaugler B, et al. Human acute myeloid leukemia CD34+/CD38– progenitor cells have decreased sensitivity to chemotherapy and Fas-induced apoptosis, reduced immunogenicity, and impaired dendritic cell transformation capacities. Cancer Res. 2000;60:4403–4411.
26. Guzman ML, Neering SJ, Upchurch D, et al. Nuclear factor-kappaB is constitutively activated in primitive human acute myelogenous leukemia cells. Blood. 2001;98:2301–2307.
27. Xu Q, Simpson SE, Scialla TJ, et al. Survival of acute myeloid leukemia cells requires PI3 kinase activation. Blood. 2003;102:972–980.
28. Xu Q, Thompson JE, Carroll M. mTOR regulates cell survival after etoposide treatment in primary AML cells. Blood. 2005;106:4261–4268.
29. Jamieson CH, Ailles LE, Dylla SJ, et al. Granulocyte-macrophage progenitors as candidate leukemic stem cells in blast-crisis CML. N Engl J Med. 2004;351:657–667.
30. Guzman ML, Rossi RM, Neelakantan S, et al. An orally bioavailable parthenolide analog selectively eradicates acute myelogenous leukemia stem and progenitor cells. Blood. 2007;110(13):4427–4435.
31. Hamilton A, Elrick L, Myssina S, et al. BCR-ABL activity and its response to drugs can be determined in CD34+ CML stem cells by CrkL phosphorylation status using flow cytometry. Leukemia. 2006;20:1035–1039.
32. Copland M, Hamilton A, Elrick LJ, et al. Dasatinib (BMS-354825) targets an earlier progenitor population than imatinib in primary CML but does not eliminate the quiescent fraction. Blood. 2006;107:4532–4539.
33. Bhatia R, Holtz M, Niu N, et al. Persistence of malignant hematopoietic progenitors in chronic myelogenous leukemia patients in complete cytogenetic remission following imatinib mesylate treatment. Blood. 2003;101:4701–4707.
34. Holtz MS, Bhatia R. Effect of imatinib mesylate on chronic myelogenous leukemia hematopoietic progenitor cells. Leuk Lymphoma. 2004;45:237–245.
35. Black JH, McCubrey JA, Willingham MC, et al. Diphtheria toxin-interleukin-3 fusion protein (DT(388)IL3) prolongs disease-free survival of leukemic immunocompromised mice. Leukemia. 2003;17:155–159.
36. Urieto JO, Liu T, Black JH, et al. Expression and purification of the recombinant diphtheria fusion toxin DT388IL3 for phase I clinical trials. Protein Expr Purif. 2004;33:123–133.
37. Yalcintepe L, Frankel AE, Hogge DE. Expression of interleukin-3 receptor subunits on defined subpopulations of acute myeloid leukemia blasts predicts the cytotoxicity of diphtheria toxin interleukin-3 fusion protein against malignant progenitors that engraft in immunodeficient mice. Blood. 2006;108:3530–3537.
38. Hogge DE, Yalcintepe L, Wong SH, et al. Variant diphtheria toxin-interleukin-3 fusion proteins with increased receptor affinity have enhanced cytotoxicity against acute myeloid leukemia progenitors. Clin Cancer Res. 2006;12:1284–1291.
39. Frankel A, Liu JS, Rizzieri D, et al. Phase I clinical study of diphtheria toxin-interleukin 3 fusion protein in patients with acute myeloid leukemia and myelodysplasia. Leuk Lymphoma. 2008;49:543–553.
40. Taussig DC, Pearce DJ, Simpson C, et al. Hematopoietic stem cells express multiple myeloid markers: implications for the origin and targeted therapy of acute myeloid leukemia. Blood. 2005;106:4086–4092.
41. Hauswirth AW, Florian S, Printz D, et al. Expression of the target receptor CD33 in CD34+/CD38–/CD123+ AML stem cells. Eur J Clin Invest. 2007;37:73–82.

42. Vercauteren S, Zapf R, Sutherland H. Primitive AML progenitors from most CD34(+) patients lack CD33 expression but progenitors from many CD34(−) AML patients express CD33. Cytotherapy. 2007;9:194–204.

43. Stasi R. Gemtuzumab ozogamicin: an anti-CD33 immunoconjugate for the treatment of acute myeloid leukaemia. Expert Opin Biol Ther. 2008;8:527–540.

44. Jin L, Hope KJ, Zhai Q, et al. Targeting of CD44 eradicates human acute myeloid leukemic stem cells. Nat Med. 2006;12:1167–1174.

45. Krause DS, Lazarides K, von Andrian UH, et al. Requirement for CD44 in homing and engraftment of BCR-ABL-expressing leukemic stem cells. Nat Med. 2006;12:1175–1180.

46. Peled A, Petit I, Kollet O, et al. Dependence of human stem cell engraftment and repopulation of NOD/SCID mice on CXCR4. Science. 1999;283:845–848.

47. Spoo AC, Lubbert M, Wierda WG, et al. CXCR4 is a prognostic marker in acute myelogenous leukemia. Blood. 2007;109:786–791.

48. Tavor S, Petit I, Porozov S, et al. CXCR4 regulates migration and development of human acute myelogenous leukemia stem cells in transplanted NOD/SCID mice. Cancer Res. 2004;64:2817–2824.

49. Gilliland DG, Griffin JD. The roles of FLT3 in hematopoiesis and leukemia. Blood. 2002;100:1532–1542.

50. Meshinchi S, Woods WG, Stirewalt DL, et al. Prevalence and prognostic significance of Flt3 internal tandem duplication in pediatric acute myeloid leukemia. Blood. 2001;97:89–94.

51. Kottaridis PD, Gale RE, Frew ME, et al. The presence of a FLT3 internal tandem duplication in patients with acute myeloid leukemia (AML) adds important prognostic information to cytogenetic risk group and response to the first cycle of chemotherapy: analysis of 854 patients from the United Kingdom Medical Research Council AML 10 and 12 trials. Blood. 2001;98:1752–1759.

52. Kiyoi H, Naoe T, Nakano Y, et al. Prognostic implication of FLT3 and N-RAS gene mutations in acute myeloid leukemia. Blood. 1999;93:3074–3080.

53. Kiyoi H, Towatari M, Yokota S, et al. Internal tandem duplication of the FLT3 gene is a novel modality of elongation mutation which causes constitutive activation of the product. Leukemia. 1998;12:1333–1337.

54. Levis M, Murphy KM, Pham R, et al. Internal tandem duplications of the FLT3 gene are present in leukemia stem cells. Blood. 2005;106:673–680.

55. Zhang W, Konopleva M, Shi YX, et al. Mutant FLT3: a direct target of sorafenib in acute myelogenous leukemia. J Natl Cancer Inst. 2008;100:184–198.

56. Corvera S, Czech MP. Direct targets of phosphoinositide 3-kinase products in membrane traffic and signal transduction. Trends Cell Biol. 1998;8:442–446.

57. Hennessy BT, Smith DL, Ram PT, et al. Exploiting the PI3K/AKT pathway for cancer drug discovery. Nat Rev Drug Discov. 2005;4:988–1004.

58. Mayo MW, Baldwin AS. The transcription factor NF-kappaB: control of oncogenesis and cancer therapy resistance. Biochim Biophys Acta. 2000;1470:M55–62.

59. Baldwin AS. Control of oncogenesis and cancer therapy resistance by the transcription factor NF-kappaB. J Clin Invest. 2001;107:241–246.

60. Bargou RC, Emmerich F, Krappmann D, et al. Constitutive nuclear factor-kappaB-RelA activation is required for proliferation and survival of Hodgkin's disease tumor cells. J Clin Invest. 1997;100:2961–2969.

61. Sweeney C, Li L, Shanmugam R, et al. Nuclear factor-kappaB is constitutively activated in prostate cancer in vitro and is overexpressed in prostatic intraepithelial neoplasia and adenocarcinoma of the prostate. Clin Cancer Res. 2004;10:5501–5507.

62. Nakshatri H, Bhat-Nakshatri P, Martin DA, et al. Constitutive activation of NF-kappaB during progression of breast cancer to hormone-independent growth. Mol Cell Biol. 1997;17:3629–3639.

63. Kim JY, Lee S, Hwangbo B, et al. NF-kappaB activation is related to the resistance of lung cancer cells to TNF-alpha-induced apoptosis. Biochem Biophys Res Commun. 2000;273:140–146.

64. Guzman ML, Rossi RM, Karnischky L, et al. The sesquiterpene lactone parthenolide induces apoptosis of human acute myelogenous leukemia stem and progenitor cells. Blood. 2005;105:4163–4169.

65. Letai AG. Diagnosing and exploiting cancer's addiction to blocks in apoptosis. Nat Rev Cancer. 2008;8:121–132.

66. Cory S, Adams JM. Killing cancer cells by flipping the Bcl-2/Bax switch. Cancer Cell. 2005;8:5–6.

67. Konopleva M, Contractor R, Tsao T, et al. Mechanisms of apoptosis sensitivity and resistance to the BH3 mimetic ABT-737 in acute myeloid leukemia. Cancer Cell. 2006;10:375–388.

68. Martinez A, Alonso M, Castro A, et al. First non-ATP competitive glycogen synthase kinase 3 beta (GSK-3beta) inhibitors: thiadiazolidinones (TDZD) as potential drugs for the treatment of Alzheimer's disease. J Med Chem. 2002;45:1292–1299.

69. Guzman ML, Li X, Corbett CA, et al. Rapid and selective death of leukemia stem and progenitor cells induced by the compound 4-benzyl, 2-methyl, 1,2,4-thiadiazolidine, 3,5 dione (TDZD-8). Blood. 2007;110(13):4436–4444.

70. Hassane DC, Guzman ML, Corbett C, et al. Discovery of agents that eradicate leukemia stem cells using an in silico screen of public gene expression data. Blood. 2008;111(12):5654–5662.

71. Klaus A, Birchmeier W. Wnt signalling and its impact on development and cancer. Nat Rev Cancer. 2008;8:387–398.

72. Hurlbut GD, Kankel MW, Lake RJ, et al. Crossing paths with Notch in the hyper-network. Curr Opin Cell Biol. 2007;19:166–175.

73. Artavanis-Tsakonas S, Rand MD, Lake RJ. Notch signaling: cell fate control and signal integration in development. Science. 1999;284:770–776.

74. Peacock CD, Wang Q, Gesell GS, et al. Hedgehog signaling maintains a tumor stem cell compartment in multiple myeloma. Proc Natl Acad Sci USA. 2007;104:4048–4053.

75. Lauth M, Toftgard R. The Hedgehog pathway as a drug target in cancer therapy. Curr Opin Invest Drugs. 2007;8:457–461.

76. Gal H, Amariglio N, Trakhtenbrot L, et al. Gene expression profiles of AML derived stem cells; similarity to hematopoietic stem cells. Leukemia. 2006;20:2147–2154.

77. van Rhenen A, Moshaver B, Kelder A, et al. Aberrant marker expression patterns on the CD34+CD38– stem cell compartment in acute myeloid leukemia allows to distinguish the malignant from the normal stem cell compartment both at diagnosis and in remission. Leukemia. 2007;21:1700–1707.

78. Milojkovic D, Devereux S, Westwood NB, et al. Antiapoptotic microenvironment of acute myeloid leukemia. J Immunol. 2004;173:6745–6752.

79. San Miguel JF, Vidriales MB, Lopez-Berges C, et al. Early immunophenotypical evaluation of minimal residual disease in acute myeloid leukemia identifies different patient risk groups and may contribute to postinduction treatment stratification. Blood. 2001;98:1746–1751.

80. Venditti A, Buccisano F, Del Poeta G, et al. Level of minimal residual disease after consolidation therapy predicts outcome in acute myeloid leukemia. Blood. 2000;96:3948–3952.

81. Kern W, Voskova D, Schoch C, et al. Determination of relapse risk based on assessment of minimal residual disease during complete remission by multiparameter flow cytometry in unselected patients with acute myeloid leukemia. Blood. 2004;104:3078–3085.

82. Feller N, van der Pol MA, van Stijn A, et al. MRD parameters using immunophenotypic detection methods are highly reliable in predicting survival in acute myeloid leukaemia. Leukemia. 2004;18:1380–1390.

83. van Rhenen A, Feller N, Kelder A, et al. High stem cell frequency in acute myeloid leukemia at diagnosis predicts high minimal residual disease and poor survival. Clin Cancer Res. 2005;11:6520–6527.

84. Goodell MA, Rosenzweig M, Kim H, et al. Dye efflux studies suggest that hematopoietic stem cells expressing low or undetectable levels of CD34 antigen exist in multiple species. Nat Med. 1997;3:1337–1345.

85. Feuring-Buske M, Hogge DE. Hoechst 33342 efflux identifies a subpopulation of cytogenetically normal CD34(+)CD38(−) progenitor cells from patients with acute myeloid leukemia. Blood. 2001;97:3882–3889.
86. Wulf GG, Wang RY, Kuehnle I, et al. A leukemic stem cell with intrinsic drug efflux capacity in acute myeloid leukemia. Blood. 2001;98:1166–1173.
87. Cox CV, Martin HM, Kearns PR, et al. Characterization of a progenitor cell population in childhood T-cell acute lymphoblastic leukemia. Blood. 2007;109:674–682.
88. Quijano CA, Moore D, Arthur D, et al. Cytogenetically aberrant cells are present in the CD34+CD33−CD38−CD19− marrow compartment in children with acute lymphoblastic leukemia. Leukemia. 1997;11:1508–1515.
89. George AA, Franklin J, Kerkof K, et al. Detection of leukemic cells in the CD34(+)CD38(−) bone marrow progenitor population in children with acute lymphoblastic leukemia. Blood. 2001;97:3925–3930.
90. Cox CV, Evely RS, Oakhill A, et al. Characterization of acute lymphoblastic leukemia progenitor cells. Blood. 2004;104:2919–2925.
91. Cox CV, Diamanti P, Evely RS, et al. Expression of CD133 in leukemia-initiating cells in childhood ALL. Blood. 2009;113:3287–3297.

SECTION III: TARGETING CANCER STEM CELL PATHWAYS

7 Hedgehog/GLI signaling in cancer

Fritz Aberger

University of Salzburg, Austria

Cancer is among the leading causes of death worldwide, and its incidence is continuously on the rise. Successful therapies of this disease are a major challenge and aim of this century. Understanding the molecular programs that control the malignant behavior of tumor cells, particularly of those that account for tumor initiation, growth, and metastasis, will be key to the development of targeted cancer therapies.

Cancer arises through the accumulation of genetic and epigenetic alterations that gradually endow the tumor cells with more aggressive growth properties, eventually leading to the spreading of cancer cells to form metastases. Over the past years, numerous studies have provided compelling evidence that many malignancies are driven by cancer stem cells, a small subpopulation of tumor cells with self-renewal and tumor-initiating capacity. Targeting the molecular signals that control self-renewal, survival, and proliferation of cancer stem cells is therefore considered a highly promising approach to tackle cancer at its very roots.[1–3]

A series of recent studies has implicated the Hedgehog/GLI (HH/GLI) signaling cascade in the development of a variety of human malignancies, and there is increasing evidence that this developmental pathway plays a critical role in cancer stem cells, making it a primary target for novel and efficient cancer therapies.[4–6] This review will give an overview on recent insights into the mechanisms of Hedgehog signal transduction, summarize key findings about the involvement of HH/GLI signaling in cancer development, and finally, concentrate on the role of HH/GLI in stem and cancer stem cells and its relevance to potential future therapies.

HEDGEHOG/GLI SIGNAL TRANSDUCTION

HH/GLI signaling fulfills a variety of critical tasks during embryonic development, including the control of pattern formation, cell differentiation, proliferation, and survival.[7] In adult organisms, HH/GLI signaling plays a fundamental role in the control of progenitor and stem cell behavior to ensure proper tissue homeostasis, regeneration, and repair.[8–12] Exquisite regulation of pathway activity in space and time is essential for such processes.[13] Spatial and temporal signal duration and strength are controlled by local production of biologically active HH ligand, which can bind to target cells, and importantly, also by the deployment of several negative feedback systems that restrain and terminate the signal at different levels of the HH/GLI cascade.[14,15] Failure to terminate the HH/GLI signaling system can have fatal consequences and is frequently associated with cancer development (see Chapter 8).

A characteristic of the HH/GLI pathway is that, in the absence of ligand stimulation, the pathway is actively repressed, mainly by the HH ligand receptor Patched (PTCH), a 12-pass transmembrane domain protein. The major function of PTCH in the absence of ligand is to prevent the activation of Smoothened (SMOH), a seven-pass transmembrane protein and essential HH effector with homology to G-protein–coupled receptors.[7,15–17] Although the mechanism of PTCH-mediated repression of SMOH is not clearly understood at present, there is evidence that negative regulation of SMOH requires a catalytic small-molecule transporter function of PTCH. In fact, PTCH shares some homology with bacterial Resistance, Nodulation, Division (RND) family members that act as small-molecule pumps. According to the current model, PTCH may function as a molecular pump to change the concentration of molecules that regulate SMOH function.[18]

The seminal discovery that mutations in intraflagellar transport (IFT) proteins affect HH/GLI signaling has greatly enhanced our current understanding of how, and, particularly, where, HH/GLI signal transduction is centered and coordinated in vertebrate cells. IFT proteins are components of the primary cilium, a solitary organelle that, like an antenna, projects from the cell surface and serves as a signal reception structure. IFT proteins regulate the transport of proteins within the primary cilium. Many of the HH pathway components, including PTCH, SMOH, and the GLI zinc finger transcription factors, are localized in the primary cilium, and it appears that IFT proteins serve to bring together the respective HH pathway components to allow signal transduction from the cell surface to the nucleus[19–22] (Figure 7–1). In fact, binding of HH protein to its receptor PTCH has been shown to remove PTCH from the primary cilium, thereby allowing SMOH to enter the cilium, where it is supposed to activate the GLI zinc finger transcription factors that control gene expression, according to pathway activity.[23,24] The entrance of SMOH into the cilium, and thus activation of HH signaling, can be induced also by the administration of SMOH agonists such as Smoothened agonist (SAG) or certain oxysterols (OS).[24] Activation of the latent GLI zinc finger transcription factors GLI2 and GLI3 appears to involve at least two critical steps: (1) prevention of the formation of C-terminally truncated GLI repressor

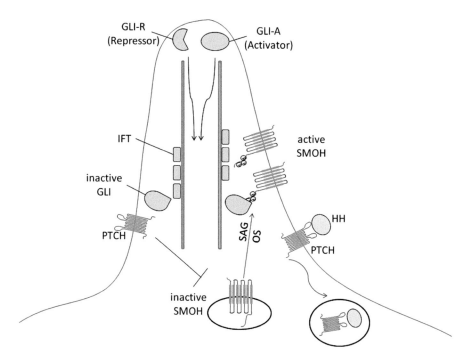

Figure 7–1: Model of HH/GLI signaling with focus on the primary cilium as the organization and coordination platform for signal transduction. GLI-R, GLI repressor form; GLI-A, GLI activator form; SAG, Smoothened agonist; OS, oxysterols; IFT, intraflagellar transport proteins. In the absence of HH ligand protein, PTCH represses SMOH and prevents its entry into the cilium. In this situation, excess GLI-R is generated and HH target genes are shut off. Binding of HH to PTCH removes PTCH from the cilium, thereby allowing SMOH to localize in the cilium and promote the formation of GLI-A forms. GLI activator forms translocate to the nucleus to induce HH target gene expression. Treatment of HH-responsive cells with SMOH agonists, such as SAG or OS, enhances the entry of SMOH into the cilium and activates signal transduction in the absence of ligand, respectively.

forms and (2) release from the negative regulation by Suppressor of Fused (SUFU), which directly interacts with GLI proteins to prevent their nuclear localization.[7,25–31] Despite some evidence that SMOH may couple and activate G proteins,[32] the details of how SMOH controls GLI activation in the primary cilium are not well understood, and it is still under debate whether the level of GLI proteins localized in the primary cilium is of sufficient physiological relevance. As GLI proteins, SUFU, and SMOH are also located outside the primary cilium, alternative cilia-independent mechanisms of signal transduction are conceivable, which may yield cellular responses different from the cilia-dependent pathway.

The vertebrate GLI proteins GLI1, GLI2, and GLI3 belong to the Ci/Gli family of zinc finger transcription factors. GLI proteins act at the distal end of the HH pathway to control gene expression. Intriguingly, GLI3, and apparently also GLI2 – but not GLI1, which has an activator function only and is mainly controlled at the level of transcription – exert a dual function as transcriptional activator and repressor.[30,33–35] In the absence of pathway activity, C-terminally truncated GLI repressor forms (GLI-R) are generated by proteolytic cleavage, which requires phosphorylation by the serine/threonine kinases Protein Kinase A (PKA),

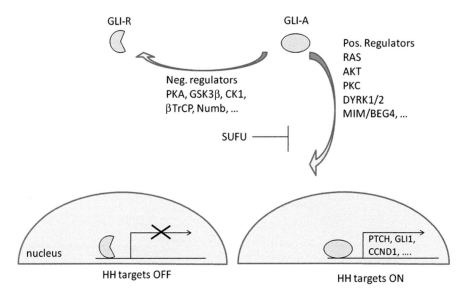

Figure 7-2: Positive and negative regulation of GLI protein activity. Negative regulators, such as PKA, GSK3β, or CK1, promote the formation of GLI repressor forms (GLI-R), which shut off transcription of HH target genes. Alternatively, βTrCP and Numb decrease the overall protein stability by enhancing the degradation of GLI proteins via the proteasome. The mechanism of action of positive regulators is less well understood. RAS and AKT may cooperate with GLI by increasing the nuclear import of GLI-A forms, which is (refers to nuclear import) normally antagonized by SUFU. Nuclear translocation of GLI-A results in the activation of HH target genes in response to pathway activation.

Glycogen Synthase Kinase 3 beta (GSK3β), and Casein Kinase I (CKI).[36–39] In addition to cytoplasmic sequestration of GLI by SUFU, the negative regulation of GLI function also involves direct interactions with βTrCP and/or Numb, which target GLI proteins to the proteosomal degradation machinery either to enhance GLI3 repressor formation or to generally reduce GLI protein stability[29,40–42] (Figure 7–2). Pathway activation through binding of HH to PTCH allows SMOH to become active, possibly by translocation to the primary cilium, thereby promoting the formation of transcriptional activator forms of the GLIs (GLI-A).

Aside from ligand-induced pathway activation, GLI protein activity can be stimulated also by interactions with other regulatory factors, including CBP, MIM/BEG4, and Dyrk1/2[43–46] (for review, see Kasper and colleagues[33]). Furthermore, an increasing body of evidence suggests that activation of GLI proteins can be promoted by pathways frequently activated in human cancers, including phosphoinositide-3 kinase (PI3K)/AKT, RAS/MEK/Extracellular Signal-Regulated Kinase (ERK), Protein Kinase C δ, and Transforming Growth Factor β/SMAD.[47–53] These signals enhance GLI activity by affecting the stability, subcellular localization, or expression of GLI proteins (reviewed in Riobo and colleagues[54] and Ruiz I Altaba and colleagues[6]) (Figure 7–2). Notably, HH/GLI and oncogenic Ras synergistically interact in the development of advanced pancreatic cancer and in melanoma growth,[53,55] and several recent studies have implicated

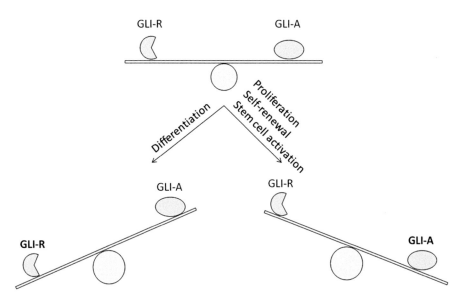

Figure 7–3: Fine-tuning of cell fate by balancing the levels of GLI repressor and GLI activator forms. In many situations, an increase of GLI-A levels is thought to induce proliferation, to activate quiescent stem cells, and to contribute to the self-renewal of stem cells. Tissue homeostasis is ensured by the reversible activation and inactivation of HH signaling and GLI-A and GLI-R formation, respectively.

the Epidermal Growth Factor Receptor (EGFR) pathway in the modulation of HH/GLI activity. EGF and Sonic HH synergistically promote neural stem cell proliferation and invasive growth of keratinocytes.[11,56,57] Our own data suggest that EGFR signaling synergizes with GLI1 and GLI2 to regulate a subset of direct GLI target genes via stimulation of RAS/RAF/MEK/ERK signaling, providing insight into the molecular mechanism that may underlie the integration of these two key oncogenic pathways. The potential pathophysiological relevance of the synergistic interaction of HH/GLI and EGFR signaling has been suggested as combined inhibition of HH/GLI and EGFR signaling efficiently decreases proliferation and survival of prostate cancer cells in vitro.[58]

HEDGEHOG/GLI SIGNALING AND CANCER DEVELOPMENT

Aside from the control of pattern formation, specification, and cell differentiation, HH/GLI plays a prominent role in the control of cell proliferation, survival, and stem cell renewal. In many settings, these processes are associated with increased pathway activity and a high GLI-A to GLI-R ratio, respectively (Figure 7–3). Given the crucial control function of HH/GLI signaling in tissue growth, repair, and regeneration, it is obvious that a precise regulation of pathway activation and termination is mandatory for the maintenance of tissue homeostasis under normal healthy conditions. This is supported by a wealth of data showing that inappropriate pathway activation by either loss of repressor function or persistent gain of pathway activator function is an etiologic factor in many human malignancies.

The involvement of HH/GLI signaling in cancer development was first discovered by the genetic analysis of Gorlin syndrome patients, who are predisposed to an early onset of multiple basal cell carcinomas (BCCs), a common non-melanoma skin cancer with low metastatic potential.[59] Gorlin patients are also at an increased risk of developing rhabdomyosarcoma and medulloblastoma, a malignant tumor of the cerebellum.[60-62] Gorlin syndrome is caused by haploinsufficiency of the HH receptor and pathway repressor PTCH. Intriguingly, hereditary as well as the majority of sporadic BCCs are characterized by loss of heterozygosity of PTCH, and about 10% of sporadic BCCs display activating mutations in the HH effector SMOH,[63-66] suggesting that ligand-independent, constitutive HH/GLI pathway activation accounts for the development of Gorlin-type cancers. This has been validated by several transgenic mouse models that allow constitutive activation of HH signaling in the epidermis, brain, or muscle.[67-76] HH/GLI signaling has recently also been identified as a critical molecular signal for the progression and maintenance of common malignant tumors with high medical need such as melanoma, lung, pancreatic, breast, and prostate cancer. As opposed to the ligand-independent Gorlin-type tumors, these tumor entities are referred to as ligand-dependent as cancer cell growth depends on autocrine or juxtacrine HH signaling.[53,55,77-84] The wide spectrum of common and fatal malignancies involving HH/GLI signaling suggests that selective pathway inhibition has a high potential as efficient antitumor strategy. This is underlined by several reports demonstrating that tumor load was dramatically reduced in vivo when mice with HH-associated cancers were treated with specific pathway inhibitors.[53,85-87] Along the same line of evidence, the administration of pathway antagonists to mice with xenografts of human lung, melanoma, glioma, pancreatic, or prostate cancer cells significantly slowed down tumor growth or, in some settings, even eliminated the tumors.[77,79,82,83,88,89] Notably, a first small clinical study with topical administration of the HH pathway inhibitor cyclopamine to BCC lesions rapidly induced tumor cell differentiation and apoptosis,[90] though results from this pilot study have to be validated in larger clinical trials.

HEDGEHOG/GLI SIGNALING IN STEM AND CANCER STEM CELLS

Seminal work on the identification and characterization of cancer stem cells, a minute subpopulation of tumor cells with self-renewal and tumor-initiating capacity, has significantly changed our current understanding of how malignant growth is regulated and where tumors originate. A wealth of recent studies has provided compelling evidence that cancer is likely to arise from stem cells or early progenitor cells and that tumor growth and disease relapse are mainly driven by cancer stem cells (reviewed in Ailles and Weissman,[1] Clarke and Fuller,[2] and Wang and Dick[3]). This concept has raised hope that targeting the cancer stem cell subpopulation may be a promising and significant step toward the cure of cancer.[91]

Several embryonic signaling pathways, including the Wnt, Notch, and HH/GLI cascades, have been implicated in stem and cancer stem cell activation and thus represent prime molecular targets for drug-based cancer therapy.[92] Under normal, healthy conditions, the HH/GLI pathway is highly active during embryonic development, in which HH signaling controls numerous developmental processes. By contrast, HH signaling in the adult organism is largely kept silent and appears to be highly restricted to settings that involve regulation of stem cell/precursor numbers or require activation of stem cells for tissue replenishment, regeneration, and repair processes. For instance, quiescent neural stem cells of the postnatal mammalian brain have been identified as targets of HH signaling, which expands this population to continuously provide new neurons.[8,11] Similarly, HH/GLI signaling controls the number of granule precursor cells of the cerebellum and the proliferation and activation of hematopoietic stem cells.[93–98] Furthermore, hormone-induced regeneration of rodent prostate tissue depends on HH/GLI pathway activation,[4,79] suggesting that putative prostate stem cells or progenitors required HH signaling to regrow this organ, although this hypothesis needs further study. There is also evidence for a stimulating function of HH/GLI on stem cells of the hair follicle. Activation of the HH pathway using a SMOH agonist has been shown to induce hair growth, possibly by activating stem cells in the resting follicle.[99] Of note, follicular stem cells have recently been shown to express HH target genes during growth of the hair follicle.[100]

The regulatory role of HH/GLI in mammalian stem cells fits a cancer model in which genetic or epigenetic changes in long-lived stem cells cause persistent activation of HH/GLI signaling and aberrant expansion of early cancer stem cells. Alternatively, such mutations may also reprogram early progenitor cells to acquire characteristics of (cancer) stem cells (i.e., the ability to self-renew, to generate different cell types, and to initiate tumor development). Both scenarios may be associated with increased HH/GLI signaling and expansion of the cancer stem cell pool. A larger pool of proliferative early precancer stem cells would be more susceptible to the accumulation of additional oncogenic mutations that may then transform the precancer stem cell into a malignant cancer stem cell (Figure 7–4). Such a process may be accelerated by the synergistic interaction of HH/GLI and other key oncogenic signals frequently activated in human cancers such as RAS, AKT, IGF, or EGFR signaling.[4,6]

Experimental evidence for an etiological function of HH/GLI signaling in (cancer) stem cells has come from several studies of different cancer types. Medulloblastoma, an embryonal tumor of the central nervous system, is probably the best studied example of a malignancy that may originate from stem or precursor cells in response to persistent HH/GLI signaling. The stem cell origin of medulloblastoma is supported by the fact that the majority of cancer cells display an undifferentiated stem cell/progenitor-like phenotype, with a few tumor cells showing signs of differentiation into neuronal, glial, and other cell types.[101,102] Activation of HH/GLI signaling has been implicated in the

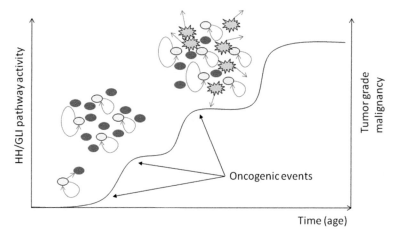

Figure 7–4: Model of cancer development and malignant growth through persistent and increasing activation of HH/GLI signaling over time. According to the model, genetic or epigenetic alterations in stem cells (light gray circles) induce persistent HH/GLI signaling that leads to proliferation and expansion of precancer stem cells (dark gray circles). The accumulation of additional mutations may activate oncogenes such as RAS or PI3K, which interact with HH/GLI signaling and further increase the level of HH/GLI activity. Persistent high-level HH/GLI signaling eventually results in the formation of metastatic cells (gray star symbols) and malignant tumor growth, respectively. Also see Ruiz I Altaba and colleagues.[6]

development of medulloblastoma in humans and mice.[69] During normal development, Sonic hedgehog (Shh) is produced by Purkinje cells and regulates the proliferation of precursor cells in the external germinal/granular layer (EGL) of the cerebellum. This has led to the hypothesis that the cellular origin of medulloblastoma may be an EGL precursor cell or even a stem cell with aberrant activation of HH/GLI signaling due to genetic loss of HH pathway repressors such as PTCH, SUFU, or REN(KCDT11).[94,95,97,98,103,104] Evidence in support of a stem cell or precursor origin comes from studies using transgenic mice that allow conditional activation of HH/GLI signaling either in the EGL or the stem cell niche of the cerebellum. Persistent activation of HH/GLI in neural stem cells of the cerebellar ventricular zone led to stem cell expansion, yet nongranule neurons or glial cells were unaffected by HH pathway activation. Only stem cell progeny that followed the granule cell lineage proliferated rapidly and gave rise to tumors with 100% penetrance, suggesting that medulloblastoma may be initiated in the stem cell. Medulloblastoma formation, however, apparently requires preceding commitment to the granule lineage.[105–107] Intriguingly, the growth of medulloblastoma can be inhibited efficiently by targeted interference with HH signaling. Also, the treatment of medulloblastoma cancer cells with HH antagonists in vitro induced molecular markers of differentiation, suggesting that HH/GLI signaling may not only promote proliferation of cancer cells, but also keep them in an undifferentiated precursor or cancer stem cell–like state.[86,97,108]

Glioblastoma, particularly glioblastoma multiforme, is a highly aggressive brain tumor. Cancer stem cells have been identified in this tumor type, and recent evidence has underlined the critical role of HH/GLI signaling in the self-renewal and tumor-initiating capacity of such cancer stem cells. Glioma cancer stem cells display activated HH/GLI signaling at the level of SMOH or further upstream.[88,109,110] Consistent with the idea that HH/GLI controls stem cell behavior via regulation of a stemness gene expression signature, interfering with HH/GLI activity was rapidly followed by a decrease in expression of key stem cell factors, including nanog, sox2, oct4, and nestin. Furthermore, inhibition of HH signaling in human glioma xenografts significantly reduced tumor growth and led to a prolonged survival of the recipient animals.[88,110] Whether the therapeutic efficacy of HH pathway inhibition is due to targeting cancer stem cells is, however, still unclear.

Support for the involvement of HH/GLI signaling in the maintenance of a cancer stem cell niche in hematological malignancies has come from studies of multiple myeloma (MM), a basically incurable cancer that arises from the clonal expansion of neoplastic plasma cells. Putative MM cancer stem cells have been characterized as a small CD138neg subpopulation that shows self-renewal and high-level HH/GLI signaling when compared to the CD138pos tumor mass.[111] Selective inhibition of HH signaling in putative MM cancer stem cells strongly reduced clonal growth and induced differentiation, while leaving the malignant growth properties of the CD138pos MM population relatively intact.[112] Targeted inhibition of HH signaling may therefore be a promising therapeutic approach toward a possible cure of MM by selectively interfering with MM cancer stem cells that are likely to account for resistance to current therapy programs and disease relapse.

Another intriguing observation is that putative mammary stem/progenitor cells express high levels of several HH pathway components, which are downregulated on differentiation. HH pathway activation in mammosphere-forming stem cells by GLI2 overexpression was shown to increase the number of putative mammary stem/progenitor cells as well as the size of the stem cell–derived mammospheres. The increase in stem/progenitor cells was dependent on GLI2-mediated induction of the polycomb factor Bmi-1, which is a key regulator of stem cell self-renewal. Notably, tumor-initiating breast cancer stem cells, much like mammary stem cells, express significantly elevated levels of HH/GLI pathway components and high levels of Bmi-1 when compared to non–tumor-initiating breast cancer cells,[113] suggesting that HH/GLI is likely to be involved in the expansion and self-renewal of breast cancer stem cells, possibly via activation of Bmi-1 expression. Whether selective blockade of the HH/GLI signaling cascade is able to inhibit breast tumor growth by affecting the cancer stem cell population still awaits experimental validation.

This also applies to the involvement of HH/GLI signaling in putative prostate and lung cancer stem cells. Although activation of the pathway has been implicated in the regeneration of prostate tissue and repair of damaged lung epithelium,[79,83] a clear demonstration of a contribution of the HH/GLI pathway to the activation and generation or expansion of stem and cancer stem cells, respectively, is still missing.

One of the first cancers that could unambiguously be ascribed to persistent and irreversible activation of HH/GLI signaling is BCC, a very common semimalignant nonmelanoma skin cancer. Support for a possible stem/progenitor cellular origin of BCC was provided using a conditional mouse model of BCC development that allows reversible activation of Gli2 expression in the epidermis and hair follicle – including the stem cell niche.[114] In this study, the authors could show that Gli-A function (here Gli2) is sufficient, and also required, for BCC growth and maintenance. Notably, inactivation of transgenic Gli2 expression in fully established Gli2-induced tumors led to rapid tumor regression, leaving residual tumors cells that were able to generate multiple epidermal lineages. The multipotency of the remaining tumor cells and their regrowth to tumors in response to Gli2-A reactivation suggest a possible epidermal/hair follicle stem cell origin of BCC, at least in this transgenic model. Whether human BCCs derive from follicular stem cells and whether HH/GLI affects growth and survival of putative BCC cancer stem cells is unclear at present, and it will be a challenging task to address these important questions.

INHIBITION OF HEDGEHOG/GLI SIGNALING IN CANCER THERAPY

During the past decade, the power of genetic and molecular biology approaches has revolutionized our current understanding of how cancers develop and how molecular signals govern tumor initiation, growth, and metastasis. More recently, it has become obvious that tumorigenesis frequently involves the redeployment of key embryonic signal cascades such as the Wnt, Notch, and HH/GLI pathways. The wealth of recent studies that have identified the HH/GLI signal as a critical molecular cue for the growth and maintenance of a great variety of malignancies, together with the emerging role of HH/GLI in cancer stem cell self-renewal, proliferation, and survival, makes this pathway a prime target for drug-based cancer therapy. Several strategies of how to inhibit HH/GLI signaling in ligand-dependent and -independent cancers are outlined in Figure 7–5.

Numerous studies have shown that targeting the function of the seven-pass transmembrane protein SMOH, which is absolutely required for HH signal transduction,[115,116] not only inhibits the activation of GLI-A forms in response to pathway activation, but may also efficiently halt or slow down the malignant growth of many HH-dependent cancer cells in vitro and in vivo (reviewed in Rubin and De Sauvage,[5] Lauth and Toftgard,[117] Ruiz I Altaba and colleagues,[118] and Xie[119]). The best documented antagonist of SMOH function is a naturally occurring plant steroidal alkaloid known as cyclopamine. Although the detailed molecular mechanism of HH pathway inhibition by cyclopamine is not completely understood at present, it has been shown that cyclopamine directly binds to the heptahelical bundle of SMOH, which is likely to promote the formation of inactive SMOH protein forms, possibly via induction of conformational changes.[120,121] Such changes may prevent SMOH from localizing to the primary cilium and thus inhibit signal transduction.[122] Several screens for SMOH antagonists have been performed, and distinct small-molecule compounds were identified that bind to SMOH and may act via the same or a similar mechanism as cyclopamine.[5,123–125] Some of

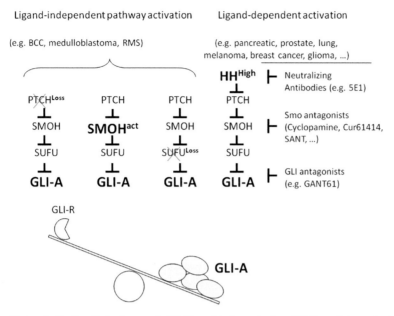

Figure 7–5: Genetic background of HH ligand-independent and HH ligand-dependent tumors and their therapeutic molecular drug targets. Gorlin-type cancers, such as BCC, medulloblastoma, or rhabdomyosarcoma (RMS), display genetic loss of PTCH or SUFU function or constitutive mutational activation of SMOH. HH ligand-dependent tumors are characterized by persistent high-level HH protein expression and constitutive autocrine pathway activation. Both HH cancer classes have in common high levels of GLI-A proteins (i.e., GLI2 and/or GLI1). Targeted inhibition of the HH pathway involves the use of ligand-neutralizing antibodies, the administration of selective SMOH antagonists such as cyclopamine, or treatment with compounds that efficiently inhibit the activity of GLI-A forms.

these compounds are currently being tested in clinical trials for their potency and usability as cancer drugs.

Another therapeutic route to selectively block pathway activation is the administration of neutralizing antibodies that bind HH protein with high affinity, thereby interfering with the binding of HH protein to its receptor PTCH.[77,83] Antibody-based approaches have proven very powerful weapons in cancer therapy. However, the use of HH-neutralizing antibodies would be limited to ligand-dependent tumors only, while ligand-independent cancers would be resistant to such treatment regimens.

The GLI transcription factors exert a critical function in response to pathway activation. GLI1 and GLI2 are considered oncogenes as their hyperactivation in response to persistent HH pathway activation drives cancer development and growth via induction of genes involved in proliferation, survival, stemness, and metastasis.[33,126–129] Selective inhibition of GLI transcription factors must therefore be considered a powerful and comprehensive strategy to antagonize HH signaling in virtually all HH-dependent tumors. In a medium-sized screen for GLI antagonists, Lauth and Toftgard[117] identified two compounds that selectively interfere with GLI function and HH-driven tumor development. The mode of action of at least one of the GLI antagonists (GANT) appears to be to disrupt or prevent binding of the GLI proteins to their target DNA sequence in promoters of target genes.[117] In addition, several naturally occurring compounds that antagonize GLI activator forms have been identified and shown to be cytotoxic to pancreatic cancer cell lines expressing HH/GLI components.[130]

As HH/GLI can cooperatively interact with other key oncogenic pathways, including EGFR, MEK/ERK, or PI3K signaling, the combined inhibition of these signals may be a highly efficient therapeutic approach. In support of this, simultaneous inhibition of EGFR and HH/GLI signaling led to synergistic reduction of prostate cancer cell viability. Similarly, concomitant interference with HH/GLI and MEK/ERK or PI3K function was shown to cooperatively decrease proliferation of melanoma cell lines in vitro.[53,58,131] More relevant in vivo studies are required to assess whether such combined approaches can hold promise for clinical applications.

Although the status of hedgehog research in the cancer field and its clinical application are still at an early stage, the ever-growing list of publications on the role of HH/GLI in malignant growth in general, and in cancer stem cells in particular, makes this signaling pathway a paradigm for how basic research on embryonic development, which started with the discovery of the fly hedgehog gene, can deepen our understanding of cancer and translate into the development of potential targeted therapies that may help cure this fatal disease.

REFERENCES

1. Ailles, L. E., and Weissman, I. L. (2007) Cancer stem cells in solid tumors. *Curr Opin Biotechnol*, **18**(5), 460–6.
2. Clarke, M. F., and Fuller, M. (2006) Stem cells and cancer: two faces of eve. *Cell*, **124**(6), 1111–5.
3. Wang, J. C., and Dick, J. E. (2005) Cancer stem cells: lessons from leukemia. *Trends Cell Biol*, **15**(9), 494–501.
4. Beachy, P. A., Karhadkar, S. S., and Berman, D. M. (2004) Tissue repair and stem cell renewal in carcinogenesis. *Nature*, **432**(7015), 324–31.
5. Rubin, L. L., and De Sauvage, F. J. (2006) Targeting the Hedgehog pathway in cancer. *Nat Rev Drug Discov*, **5**(12), 1026–33.
6. Ruiz I Altaba, A., Mas, C., and Stecca, B. (2007) The Gli code: an information nexus regulating cell fate, stemness and cancer. *Trends Cell Biol*, **17**(9), 438–47.
7. Ingham, P. W., and McMahon, A. P. (2001) Hedgehog signaling in animal development: paradigms and principles. *Genes Dev*, **15**(23), 3059–87.
8. Ahn, S., and Joyner, A. L. (2005) In vivo analysis of quiescent adult neural stem cells responding to Sonic hedgehog. *Nature*, **437**(7060), 894–7.
9. Asai, J., Takenaka, H., Kusano, K. F., Ii, M., Luedemann, C., Curry, C., Eaton, E., Iwakura, A., Tsutsumi, Y., Hamada, H., Kishimoto, S., Thorne, T., Kishore, R., and Losordo, D. W. (2006) Topical Sonic hedgehog gene therapy accelerates wound healing in diabetes by enhancing endothelial progenitor cell-mediated microvascular remodeling. *Circulation*, **113**(20), 2413–24.
10. Chiang, C., Swan, R. Z., Grachtchouk, M., Bolinger, M., Litingtung, Y., Robertson, E. K., Cooper, M. K., Gaffield, W., Westphal, H., Beachy, P. A., and Dlugosz, A. A. (1999) Essential role for Sonic hedgehog during hair follicle morphogenesis. *Dev Biol*, **205**(1), 1–9.
11. Palma, V., Lim, D. A., Dahmane, N., Sanchez, P., Brionne, T. C., Herzberg, C. D., Gitton, Y., Carleton, A., Alvarez-Buylla, A., and Ruiz I Altaba, A. (2005) Sonic hedgehog

controls stem cell behavior in the postnatal and adult brain. *Development*, **132**(2), 335–44.

12. St-Jacques, B., Dassule, H. R., Karavanova, I., Botchkarev, V. A., Li, J., Danielian, P. S., McMahon, J. A., Lewis, P. M., Paus, R., and McMahon, A. P. (1998) Sonic hedgehog signaling is essential for hair development. *Curr Biol*, **8**(19), 1058–68.

13. Oro, A. E., and Higgins, K. (2003) Hair cycle regulation of Hedgehog signal reception. *Dev Biol*, **255**(2), 238–48.

14. Jiang, J. (2006) Regulation of Hh/Gli signaling by dual ubiquitin pathways. *Cell Cycle*, **5**(21), 2457–63.

15. Wang, Y., McMahon, A. P., and Allen, B. L. (2007) Shifting paradigms in Hedgehog signaling. *Curr Opin Cell Biol*, **19**(2), 159–65.

16. Hooper, J. E., and Scott, M. P. (2005) Communicating with Hedgehogs. *Nat Rev Mol Cell Biol*, **6**(4), 306–17.

17. Varjosalo, M., Li, S. P., and Taipale, J. (2006) Divergence of Hedgehog signal transduction mechanism between Drosophila and mammals. *Dev Cell*, **10**(2), 177–86.

18. Taipale, J., Cooper, M. K., Maiti, T., and Beachy, P. A. (2002) Patched acts catalytically to suppress the activity of Smoothened. *Nature*, **418**(6900), 892–7.

19. Eggenschwiler, J. T., and Anderson, K. V. (2007) Cilia and developmental signaling. *Annu Rev Cell Dev Biol*, **23**, 345–73.

20. Huangfu, D., and Anderson, K. V. (2006) Signaling from Smo to Ci/Gli: conservation and divergence of Hedgehog pathways from Drosophila to vertebrates. *Development*, **133**(1), 3–14.

21. Oro, A. E. (2007) The primary cilia, a "rab-id" transit system for Hedgehog signaling. *Curr Opin Cell Biol*, **19**(6), 691–6.

22. Scholey, J. M., and Anderson, K. V. (2006) Intraflagellar transport and cilium-based signaling. *Cell*, **125**(3), 439–42.

23. Haycraft, C. J., Banizs, B., Aydin-Son, Y., Zhang, Q., Michaud, E. J., and Yoder, B. K. (2005) Gli2 and Gli3 localize to cilia and require the intraflagellar transport protein polaris for processing and function. *PLoS Genet*, **1**(4), e53.

24. Rohatgi, R., Milenkovic, L., and Scott, M. P. (2007) Patched1 regulates Hedgehog signaling at the primary cilium. *Science*, **317**(5836), 372–6.

25. Aza-Blanc, P., Ramirez-Weber, F. A., Laget, M. P., Schwartz, C., and Kornberg, T. B. (1997) Proteolysis that is inhibited by Hedgehog targets Cubitus interruptus protein to the nucleus and converts it to a repressor. *Cell*, **89**(7), 1043–53.

26. Kogerman, P., Grimm, T., Kogerman, L., Krause, D., Unden, A. B., Sandstedt, B., Toftgard, R., and Zaphiropoulos, P. G. (1999) Mammalian suppressor-of-Fused modulates nuclear-cytoplasmic shuttling of Gli-1. *Nat Cell Biol*, **1**(5), 312–9.

27. Methot, N., and Basler, K. (1999) Hedgehog controls limb development by regulating the activities of distinct transcriptional activator and repressor forms of Cubitus interruptus. *Cell*, **96**(6), 819–31.

28. Methot, N., and Basler, K. (2000) Suppressor of Fused opposes Hedgehog signal transduction by impeding nuclear accumulation of the activator form of Cubitus interruptus. *Development*, **127**(18), 4001–10.

29. Pan, Y., Bai, C. B., Joyner, A. L., and Wang, B. (2006) Sonic hedgehog signaling regulates Gli2 transcriptional activity by suppressing its processing and degradation. *Mol Cell Biol*, **26**(9), 3365–77.

30. Ruiz I Altaba, A. (1999) Gli proteins encode context-dependent positive and negative functions: implications for development and disease. *Development*, **126**(14), 3205–16.

31. Wang, B., Fallon, J. F., and Beachy, P. A. (2000) Hedgehog-regulated processing of Gli3 produces an anterior/posterior repressor gradient in the developing vertebrate limb. *Cell*, **100**(4), 423–34.

32. Riobo, N. A., Saucy, B., Dilizio, C., and Manning, D. R. (2006) Activation of heterotrimeric G proteins by Smoothened. *Proc Natl Acad Sci U S A*, **103**(33), 12607–12.

33. Kasper, M., Regl, G., Frischauf, A. M., and Aberger, F. (2006) Gli transcription factors: mediators of oncogenic Hedgehog signalling. *Eur J Cancer*, **42**(4), 437–45.

34. Litingtung, Y., and Chiang, C. (2000) Specification of ventral neuron types is mediated by an antagonistic interaction between Shh and Gli3. *Nat Neurosci*, **3**(10), 979–85.

35. Litingtung, Y., Dahn, R. D., Li, Y., Fallon, J. F., and Chiang, C. (2002) Shh and Gli3 are dispensable for limb skeleton formation but regulate digit number and identity. *Nature*, **418**(6901), 979–83.

36. Jia, J., Zhang, L., Zhang, Q., Tong, C., Wang, B., Hou, F., Amanai, K., and Jiang, J. (2005) Phosphorylation by double-time/CKiepsilon and CKialpha targets Cubitus interruptus for Slimb/beta-TRCP-mediated proteolytic processing. *Dev Cell*, **9**(6), 819–30.

37. Price, M. A., and Kalderon, D. (2002) Proteolysis of the Hedgehog signaling effector Cubitus interruptus requires phosphorylation by glycogen synthase kinase 3 and casein kinase 1. *Cell*, **108**(6), 823–35.

38. Tempe, D., Casas, M., Karaz, S., Blanchet-Tournier, M. F., and Concordet, J. P. (2006) Multisite protein kinase A and glycogen synthase kinase 3beta phosphorylation leads to Gli3 ubiquitination by SCFbetaTrCP. *Mol Cell Biol*, **26**(11), 4316–26.

39. Wang, B., and Li, Y. (2006) Evidence for the direct involvement of {beta}TrCP in Gli3 protein processing. *Proc Natl Acad Sci U S A*, **103**(1), 33–8.

40. Bhatia, N., Thiyagarajan, S., Elcheva, I., Saleem, M., Dlugosz, A., Mukhtar, H., and Spiegelman, V. S. (2006) Gli2 is targeted for ubiquitination and degradation by beta-TrCP ubiquitin ligase. *J Biol Chem*, **281**(28), 19320–6.

41. Di Marcotullio, L., Ferretti, E., Greco, A., De Smaele, E., Po, A., Sico, M. A., Alimandi, M., Giannini, G., Maroder, M., Screpanti, I., and Gulino, A. (2006) Numb is a suppressor of Hedgehog signalling and targets Gli1 for itch-dependent ubiquitination. *Nat Cell Biol*, **8**(12), 1415–23.

42. Huntzicker, E. G., Estay, I. S., Zhen, H., Lokteva, L. A., Jackson, P. K., and Oro, A. E. (2006) Dual degradation signals control Gli protein stability and tumor formation. *Genes Dev*, **20**(3), 276–81.

43. Callahan, C. A., Ofstad, T., Horng, L., Wang, J. K., Zhen, H. H., Coulombe, P. A., and Oro, A. E. (2004) Mim/beg4, a Sonic hedgehog-responsive gene that potentiates Gli-dependent transcription. *Genes Dev*, **18**(22), 2724–9.

44. Dai, P., Akimaru, H., Tanaka, Y., Maekawa, T., Nakafuku, M., and Ishii, S. (1999) Sonic hedgehog–induced activation of the Gli1 promoter is mediated by Gli3. *J Biol Chem*, **274**(12), 8143–52.

45. Mao, J., Maye, P., Kogerman, P., Tejedor, F. J., Toftgard, R., Xie, W., Wu, G., and Wu, D. (2002) Regulation of Gli1 transcriptional activity in the nucleus by Dyrk1. *J Biol Chem*, **277**(38), 35156–61.

46. Varjosalo, M., Bjorklund, M., Cheng, F., Syvanen, H., Kivioja, T., Kilpinen, S., Sun, Z., Kallioniemi, O., Stunnenberg, H. G., He, W. W., Ojala, P., and Taipale, J. (2008) Application of active and kinase-deficient kinome collection for identification of kinases regulating Hedgehog signaling. *Cell*, **133**(3), 537–48.

47. Dennler, S., Andre, J., Alexaki, I., Li, A., Magnaldo, T., Ten Dijke, P., Wang, X. J., Verrecchia, F., and Mauviel, A. (2007) Induction of Sonic hedgehog mediators by transforming growth factor-beta: Smad3-dependent activation of Gli2 and Gli1 expression in vitro and in vivo. *Cancer Res*, **67**(14), 6981–6.

48. Kasper, M., Schnidar, H., Neill, G. W., Hanneder, M., Klingler, S., Blaas, L., Schmid, C., Hauser-Kronberger, C., Regl, G., Philpott, M. P., and Aberger, F. (2006) Selective modulation of Hedgehog/Gli target gene expression by epidermal growth factor signaling in human keratinocytes. *Mol Cell Biol*, **26**(16), 6283–98.

49. Lauth, M., Bergstrom, A., and Toftgard, R. (2007) Phorbol esters inhibit the Hedgehog signalling pathway downstream of Suppressor of Fused, but upstream of Gli. *Oncogene*, **26**(35), 5163–8.

50. Lauth, M., and Toftgard, R. (2007) Non-canonical activation of Gli transcription factors: implications for targeted anti-cancer therapy. *Cell Cycle*, **6**(20), 2458–63.

51. Riobo, N. A., Haines, G. M., and Emerson, C. P., Jr. (2006) Protein kinase c-delta and mitogen-activated protein/extracellular signal-regulated kinase-1 control Gli activation in Hedgehog signaling. *Cancer Res*, **66**(2), 839–45.

52. Riobo, N. A., Lu, K., Ai, X., Haines, G. M., and Emerson, C. P., Jr. (2006) Phosphoinositide 3-kinase and Akt are essential for Sonic hedgehog signaling. *Proc Natl Acad Sci U S A*, **103**(12), 4505–10.

53. Stecca, B., Mas, C., Clement, V., Zbinden, M., Correa, R., Piguet, V., Beermann, F., and Ruiz, I. A. A. (2007) Melanomas require Hedgehog-Gli signaling regulated by interactions between Gli1 and the Ras-Mek/Akt pathways. *Proc Natl Acad Sci U S A*, **104**(14), 5895–900.

54. Riobo, N. A., Lu, K., and Emerson, C. P., Jr. (2006) Hedgehog signal transduction: signal integration and cross talk in development and cancer. *Cell Cycle*, **5**(15), 1612–5.

55. Pasca Di Magliano, M., Sekine, S., Ermilov, A., Ferris, J., Dlugosz, A. A., and Hebrok, M. (2006) Hedgehog/Ras interactions regulate early stages of pancreatic cancer. *Genes Dev*, **20**(22), 3161–73.

56. Bigelow, R. L., Jen, E. Y., Delehedde, M., Chari, N. S., and McDonnell, T. J. (2005) Sonic hedgehog induces epidermal growth factor dependent matrix infiltration in HaCaT keratinocytes. *J Invest Dermatol*, **124**(2), 457–65.

57. Palma, V., and Ruiz I Altaba, A. (2004) Hedgehog-Gli signaling regulates the behavior of cells with stem cell properties in the developing neocortex. *Development*, **131**(2), 337–45.

58. Mimeault, M., Moore, E., Moniaux, N., Henichart, J. P., Depreux, P., Lin, M. F., and Batra, S. K. (2006) Cytotoxic effects induced by a combination of cyclopamine and gefitinib, the selective Hedgehog and epidermal growth factor receptor signaling inhibitors, in prostate cancer cells. *Int J Cancer*, **118**(4), 1022–31.

59. Gorlin, R. J. (1995) Nevoid basal cell carcinoma syndrome. *Dermatol Clin*, **13**(1), 113–25.

60. Bale, A. E. (2002) Hedgehog signaling and human disease. *Annu Rev Genomics Hum Genet*, **3**, 47–65.

61. Gailani, M. R., Stahle-Backdahl, M., Leffell, D. J., Glynn, M., Zaphiropoulos, P. G., Pressman, C., Unden, A. B., Dean, M., Brash, D. E., Bale, A. E., and Toftgard, R. (1996) The role of the human homologue of Drosophila Patched in sporadic basal cell carcinomas. *Nat Genet*, **14**(1), 78–81.

62. Goodrich, L. V., and Scott, M. P. (1998) Hedgehog and Patched in neural development and disease. *Neuron*, **21**(6), 1243–57.

63. Hahn, H., Wojnowski, L., Miller, G., and Zimmer, A. (1999) The Patched signaling pathway in tumorigenesis and development: lessons from animal models. *J Mol Med*, **77**(6), 459–68.

64. Johnson, R. L., Rothman, A. L., Xie, J., Goodrich, L. V., Bare, J. W., Bonifas, J. M., Quinn, A. G., Myers, R. M., Cox, D. R., Epstein, E. H., Jr., and Scott, M. P. (1996) Human homolog of Patched, a candidate gene for the basal cell nevus syndrome. *Science*, **272**(5268), 1668–71.

65. Lam, C. W., Xie, J., To, K. F., Ng, H. K., Lee, K. C., Yuen, N. W., Lim, P. L., Chan, L. Y., Tong, S. F., and McCormick, F. (1999) A frequent activated Smoothened mutation in sporadic basal cell carcinomas. *Oncogene*, **18**(3), 833–6.

66. Xie, J., Murone, M., Luoh, S. M., Ryan, A., Gu, Q., Zhang, C., Bonifas, J. M., Lam, C. W., Hynes, M., Goddard, A., Rosenthal, A., Epstein, E. H., Jr., and De Sauvage, F. J. (1998) Activating Smoothened mutations in sporadic basal-cell carcinoma. *Nature*, **391**(6662), 90–2.

67. Aszterbaum, M., Epstein, J., Oro, A., Douglas, V., Leboit, P. E., Scott, M. P., and Epstein, E. H., Jr. (1999) Ultraviolet and ionizing radiation enhance the growth of BCCs and trichoblastomas in Patched heterozygous knockout mice. *Nat Med*, **5**(11), 1285–91.

68. Fan, H., Oro, A. E., Scott, M. P., and Khavari, P. A. (1997) Induction of basal cell carcinoma features in transgenic human skin expressing Sonic hedgehog. *Nat Med*, **3**(7), 788–92.

69. Goodrich, L. V., Milenkovic, L., Higgins, K. M., and Scott, M. P. (1997) Altered neural cell fates and medulloblastoma in mouse Patched mutants. *Science*, **277**(5329), 1109–13.

70. Grachtchouk, M., Mo, R., Yu, S., Zhang, X., Sasaki, H., Hui, C. C., and Dlugosz, A. A. (2000) Basal cell carcinomas in mice overexpressing Gli2 in skin. *Nat Genet*, **24**(3), 216–7.

71. Hallahan, A. R., Pritchard, J. I., Hansen, S., Benson, M., Stoeck, J., Hatton, B. A., Russell, T. L., Ellenbogen, R. G., Bernstein, I. D., Beachy, P. A., and Olson, J. M. (2004) The Smoal mouse model reveals that Notch signaling is critical for the growth and survival of Sonic hedgehog–induced medulloblastomas. *Cancer Res*, **64**(21), 7794–800.

72. Hatton, B. A., Villavicencio, E. H., Tsuchiya, K. D., Pritchard, J. I., Ditzler, S., Pullar, B., Hansen, S., Knoblaugh, S. E., Lee, D., Eberhart, C. G., Hallahan, A. R., and Olson, J. M. (2008) The Smo/Smo model: Hedgehog-induced medulloblastoma with 90% incidence and leptomeningeal spread. *Cancer Res*, **68**(6), 1768–76.

73. Kappler, R., Bauer, R., Calzada-Wack, J., Rosemann, M., Hemmerlein, B., and Hahn, H. (2004) Profiling the molecular difference between Patched- and p53-dependent rhabdomyosarcoma. *Oncogene*, **23**(54), 8785–95.

74. Nilsson, M., Unden, A. B., Krause, D., Malmqwist, U., Raza, K., Zaphiropoulos, P. G., and Toftgard, R. (2000) Induction of basal cell carcinomas and trichoepitheliomas in mice overexpressing Gli-1. *Proc Natl Acad Sci U S A*, **97**(7), 3438–43.

75. Oro, A. E., Higgins, K. M., Hu, Z., Bonifas, J. M., Epstein, E. H., Jr., and Scott, M. P. (1997) Basal cell carcinomas in mice overexpressing Sonic hedgehog. *Science*, **276**(5313), 817–21.

76. Wetmore, C., Eberhart, D. E., and Curran, T. (2000) The normal Patched allele is expressed in medulloblastomas from mice with heterozygous germ-line mutation of Patched. *Cancer Res*, **60**(8), 2239–46.

77. Berman, D. M., Karhadkar, S. S., Maitra, A., Montes De Oca, R., Gerstenblith, M. R., Briggs, K., Parker, A. R., Shimada, Y., Eshleman, J. R., Watkins, D. N., and Beachy, P. A. (2003) Widespread requirement for Hedgehog ligand stimulation in growth of digestive tract tumours. *Nature*, **425**(6960), 846–51.

78. Hatsell, S., and Frost, A. R. (2007) Hedgehog signaling in mammary gland development and breast cancer. *J Mammary Gland Biol Neoplasia*, **12**(2–3), 163–73.

79. Karhadkar, S. S., Bova, G. S., Abdallah, N., Dhara, S., Gardner, D., Maitra, A., Isaacs, J. T., Berman, D. M., and Beachy, P. A. (2004) Hedgehog signalling in prostate regeneration, neoplasia and metastasis. *Nature*, **431**(7009), 707–12.

80. Kubo, M., Nakamura, M., Tasaki, A., Yamanaka, N., Nakashima, H., Nomura, M., Kuroki, S., and Katano, M. (2004) Hedgehog signaling pathway is a new therapeutic target for patients with breast cancer. *Cancer Res*, **64**(17), 6071–4.

81. Sanchez, P., Hernandez, A. M., Stecca, B., Kahler, A. J., Degueme, A. M., Barrett, A., Beyna, M., Datta, M. W., Datta, S., and Ruiz I Altaba, A. (2004) Inhibition of prostate cancer proliferation by interference with Sonic hedgehog-Gli1 signaling. *Proc Natl Acad Sci U S A*, **101**(34), 12561–6.

82. Thayer, S. P., Di Magliano, M. P., Heiser, P. W., Nielsen, C. M., Roberts, D. J., Lauwers, G. Y., Qi, Y. P., Gysin, S., Fernandez-Del Castillo, C., Yajnik, V., Antoniu, B., McMahon, M., Warshaw, A. L., and Hebrok, M. (2003) Hedgehog is an early and late mediator of pancreatic cancer tumorigenesis. *Nature*, **425**(6960), 851–6.

83. Watkins, D. N., Berman, D. M., Burkholder, S. G., Wang, B., Beachy, P. A., and Baylin, S. B. (2003) Hedgehog signalling within airway epithelial progenitors and in small-cell lung cancer. *Nature*, **422**(6929), 313–7.

84. Yuan, Z., Goetz, J. A., Singh, S., Ogden, S. K., Petty, W. J., Black, C. C., Memoli, V. A., Dmitrovsky, E., and Robbins, D. J. (2007) Frequent requirement of Hedgehog signaling in non-small cell lung carcinoma. *Oncogene*, **26**(7), 1046–55.

85. Athar, M., Li, C., Tang, X., Chi, S., Zhang, X., Kim, A. L., Tyring, S. K., Kopelovich, L., Hebert, J., Epstein, E. H., Jr., Bickers, D. R., and Xie, J. (2004) Inhibition of Smoothened signaling prevents ultraviolet b-induced basal cell carcinomas through regulation of Fas expression and apoptosis. *Cancer Res*, **64**(20), 7545–52.

86. Romer, J. T., Kimura, H., Magdaleno, S., Sasai, K., Fuller, C., Baines, H., Connelly, M., Stewart, C. F., Gould, S., Rubin, L. L., and Curran, T. (2004) Suppression of the Shh pathway using a small molecule inhibitor eliminates medulloblastoma in Ptc1(+/−) p53(−/−) mice. *Cancer Cell*, **6**(3), 229–40.

87. Sanchez, P., and Ruiz I Altaba, A. (2005) In vivo inhibition of endogenous brain tumors through systemic interference of Hedgehog signaling in mice. *Mech Dev*, **122**(2), 223–30.

88. Clement, V., Sanchez, P., De Tribolet, N., Radovanovic, I., and Ruiz I Altaba, A. (2007) Hedgehog-Gli1 signaling regulates human glioma growth, cancer stem cell self-renewal, and tumorigenicity. *Curr Biol*, **17**(2), 165–72.

89. Lauth, M., Bergstrom, A., Shimokawa, T., and Toftgard, R. (2007) Inhibition of Gli-mediated transcription and tumor cell growth by small-molecule antagonists. *Proc Natl Acad Sci U S A*, **104**(20), 8455–60.

90. Tabs, S., and Avci, O. (2004) Induction of the differentiation and apoptosis of tumor cells in vivo with efficiency and selectivity. *Eur J Dermatol*, **14**(2), 96–102.

91. Boman, B. M., and Wicha, M. S. (2008) Cancer stem cells: a step toward the cure. *J Clin Oncol*, **26**(17), 2795–9.

92. Reya, T., Morrison, S. J., Clarke, M. F., and Weissman, I. L. (2001) Stem cells, cancer, and cancer stem cells. *Nature*, **414**(6859), 105–11.

93. Bhardwaj, G., Murdoch, B., Wu, D., Baker, D. P., Williams, K. P., Chadwick, K., Ling, L. E., Karanu, F. N., and Bhatia, M. (2001) Sonic hedgehog induces the proliferation of primitive human hematopoietic cells via BMP regulation. *Nat Immunol*, **2**(2), 172–80.

94. Dahmane, N., and Ruiz I Altaba, A. (1999) Sonic hedgehog regulates the growth and patterning of the cerebellum. *Development*, **126**(14), 3089–100.

95. Kenney, A. M., and Rowitch, D. H. (2000) Sonic hedgehog promotes G(1) cyclin expression and sustained cell cycle progression in mammalian neuronal precursors. *Mol Cell Biol*, **20**(23), 9055–67.

96. Trowbridge, J. J., Scott, M. P., and Bhatia, M. (2006) Hedgehog modulates cell cycle regulators in stem cells to control hematopoietic regeneration. *Proc Natl Acad Sci U S A*, **103**(38), 14134–9.

97. Wallace, V. A. (1999) Purkinje-cell-derived Sonic hedgehog regulates granule neuron precursor cell proliferation in the developing mouse cerebellum. *Curr Biol*, **9**(8), 445–8.

98. Wechsler-Reya, R. J., and Scott, M. P. (1999) Control of neuronal precursor proliferation in the cerebellum by Sonic hedgehog. *Neuron*, **22**(1), 103–14.

99. Paladini, R. D., Saleh, J., Qian, C., Xu, G. X., and Rubin, L. L. (2005) Modulation of hair growth with small molecule agonists of the Hedgehog signaling pathway. *J Invest Dermatol*, **125**(4), 638–46.

100. Jaks, V., Barker, N., Kasper, M., Van Es, J. H., Snippert, H. J., and Toftgard, R. (2009) Lgr5 marks cycling, yet long-lived, hair follicle stem cells. *Nat Genet*, in press.

101. Fan, X., and Eberhart, C. G. (2008) Medulloblastoma stem cells. *J Clin Oncol*, **26**(17), 2821–7.

102. Louis, D. N., Ohgaki, H., Wiestler, O. D., Cavenee, W. K., Burger, P. C., Jouvet, A., Scheithauer, B. W., and Kleihues, P. (2007) The 2007 WHO classification of tumours of the central nervous system. *Acta Neuropathol*, **114**(2), 97–109.

103. Di Marcotullio, L., Ferretti, E., De Smaele, E., Argenti, B., Mincione, C., Zazzeroni, F., Gallo, R., Masuelli, L., Napolitano, M., Maroder, M., Modesti, A., Giangaspero, F., Screpanti, I., Alesse, E., and Gulino, A. (2004) Ren(kctd11) is a suppressor of

Hedgehog signaling and is deleted in human medulloblastoma. *Proc Natl Acad Sci U S A*, **101**(29), 10833–8.

104. Taylor, M. D., Liu, L., Raffel, C., Hui, C. C., Mainprize, T. G., Zhang, X., Agatep, R., Chiappa, S., Gao, L., Lowrance, A., Hao, A., Goldstein, A. M., Stavrou, T., Scherer, S. W., Dura, W. T., Wainwright, B., Squire, J. A., Rutka, J. T., and Hogg, D. (2002) Mutations in SUFU predispose to medulloblastoma. *Nat Genet*, **31**(3), 306–10.
105. Eberhart, C. G. (2008) Even cancers want commitment: lineage identity and medulloblastoma formation. *Cancer Cell*, **14**(2), 105–7.
106. Schuller, U., Heine, V. M., Mao, J., Kho, A. T., Dillon, A. K., Han, Y. G., Huillard, E., Sun, T., Ligon, A. H., Qian, Y., Ma, Q., Alvarez-Buylla, A., McMahon, A. P., Rowitch, D. H., and Ligon, K. L. (2008) Acquisition of granule neuron precursor identity is a critical determinant of progenitor cell competence to form Shh-induced medulloblastoma. *Cancer Cell*, **14**(2), 123–34.
107. Yang, Z. J., Ellis, T., Markant, S. L., Read, T. A., Kessler, J. D., Bourboulas, M., Schuller, U., Machold, R., Fishell, G., Rowitch, D. H., Wainwright, B. J., and Wechsler-Reya, R. J. (2008) Medulloblastoma can be initiated by deletion of Patched in lineage-restricted progenitors or stem cells. *Cancer Cell*, **14**(2), 135–45.
108. Berman, D. M., Karhadkar, S. S., Hallahan, A. R., Pritchard, J. I., Eberhart, C. G., Watkins, D. N., Chen, J. K., Cooper, M. K., Taipale, J., Olson, J. M., and Beachy, P. A. (2002) Medulloblastoma growth inhibition by Hedgehog pathway blockade. *Science*, **297**(5586), 1559–61.
109. Ehtesham, M., Sarangi, A., Valadez, J. G., Chanthaphaychith, S., Becher, M. W., Abel, T. W., Thompson, R. C., and Cooper, M. K. (2007) Ligand-dependent activation of the Hedgehog pathway in glioma progenitor cells. *Oncogene*, **26**(39), 5752–61.
110. Bar, E. E., Chaudhry, A., Lin, A., Fan, X., Schreck, K., Matsui, W., Piccirillo, S., Vescovi, A. L., Dimeco, F., Olivi, A., and Eberhart, C. G. (2007) Cyclopamine-mediated Hedgehog pathway inhibition depletes stem-like cancer cells in glioblastoma. *Stem Cells*, **25**(10), 2524–33.
111. Matsui, W., Huff, C. A., Wang, Q., Malehorn, M. T., Barber, J., Tanhehco, Y., Smith, B. D., Civin, C. I., and Jones, R. J. (2004) Characterization of clonogenic multiple myeloma cells. *Blood*, **103**(6), 2332–6.
112. Peacock, C. D., Wang, Q., Gesell, G. S., Corcoran-Schwartz, I. M., Jones, E., Kim, J., Devereux, W. L., Rhodes, J. T., Huff, C. A., Beachy, P. A., Watkins, D. N., and Matsui, W. (2007) Hedgehog signaling maintains a tumor stem cell compartment in multiple myeloma. *Proc Natl Acad Sci U S A*, **104**(10), 4048–53.
113. Liu, S., Dontu, G., Mantle, I. D., Patel, S., Ahn, N. S., Jackson, K. W., Suri, P., and Wicha, M. S. (2006) Hedgehog signaling and BMI-1 regulate self-renewal of normal and malignant human mammary stem cells. *Cancer Res*, **66**(12), 6063–71.
114. Hutchin, M. E., Kariapper, M. S., Grachtchouk, M., Wang, A., Wei, L., Cummings, D., Liu, J., Michael, L. E., Glick, A., and Dlugosz, A. A. (2005) Sustained Hedgehog signaling is required for basal cell carcinoma proliferation and survival: conditional skin tumorigenesis recapitulates the hair growth cycle. *Genes Dev*, **19**(2), 214–23.
115. Van Den Heuvel, M., and Ingham, P. W. (1996) Smoothened encodes a receptor-like serpentine protein required for Hedgehog signalling. *Nature*, **382**(6591), 547–51.
116. Zhang, X. M., Ramalho-Santos, M., and McMahon, A. P. (2001) Smoothened mutants reveal redundant roles for Shh and Ihh signaling including regulation of L/R asymmetry by the mouse node. *Cell*, **105**(6), 781–92.
117. Lauth, M., and Toftgard, R. (2007) The Hedgehog pathway as a drug target in cancer therapy. *Curr Opin Invest Drugs*, **8**(6), 457–61.
118. Ruiz I Altaba, A., Sanchez, P., and Dahmane, N. (2002) Gli and Hedgehog in cancer: tumours, embryos and stem cells. *Nat Rev Cancer*, **2**(5), 361–72.
119. Xie, J. (2008) Hedgehog signaling pathway: development of antagonists for cancer therapy. *Curr Oncol Rep*, **10**(2), 107–13.

120. Chen, J. K., Taipale, J., Cooper, M. K., and Beachy, P. A. (2002) Inhibition of Hedgehog signaling by direct binding of cyclopamine to Smoothened. *Genes Dev*, **16**(21), 2743–8.

121. Taipale, J., Chen, J. K., Cooper, M. K., Wang, B., Mann, R. K., Milenkovic, L., Scott, M. P., and Beachy, P. A. (2000) Effects of oncogenic mutations in Smoothened and Patched can be reversed by cyclopamine. *Nature*, **406**(6799), 1005–9.

122. Corbit, K. C., Aanstad, P., Singla, V., Norman, A. R., Stainier, D. Y., and Reiter, J. F. (2005) Vertebrate Smoothened functions at the primary cilium. *Nature*, **437**(7061), 1018–21.

123. Chen, J. K., Taipale, J., Young, K. E., Maiti, T., and Beachy, P. A. (2002) Small molecule modulation of Smoothened activity. *Proc Natl Acad Sci U S A*, **99**(22), 14071–6.

124. Frank-Kamenetsky, M., Zhang, X. M., Bottega, S., Guicherit, O., Wichterle, H., Dudek, H., Bumcrot, D., Wang, F. Y., Jones, S., Shulok, J., Rubin, L. L., and Porter, J. A. (2002) Small-molecule modulators of Hedgehog signaling: identification and characterization of Smoothened agonists and antagonists. *J Biol*, **1**(2), 10.

125. Williams, J. A., Guicherit, O. M., Zaharian, B. I., Xu, Y., Chai, L., Wichterle, H., Kon, C., Gatchalian, C., Porter, J. A., Rubin, L. L., and Wang, F. Y. (2003) Identification of a small molecule inhibitor of the Hedgehog signaling pathway: effects on basal cell carcinoma-like lesions. *Proc Natl Acad Sci U S A*, **100**(8), 4616–21.

126. Bigelow, R. L., Chari, N. S., Unden, A. B., Spurgers, K. B., Lee, S., Roop, D. R., Toftgard, R., and Mcdonnell, T. J. (2004) Transcriptional regulation of BCL-2 mediated by the Sonic hedgehog signaling pathway through Gli-1. *J Biol Chem*, **279**(2), 1197–205.

127. Louro, I. D., Bailey, E. C., Li, X., South, L. S., McKie-Bell, P. R., Yoder, B. K., Huang, C. C., Johnson, M. R., Hill, A. E., Johnson, R. L., and Ruppert, J. M. (2002) Comparative gene expression profile analysis of Gli and c-MYC in an epithelial model of malignant transformation. *Cancer Res*, **62**(20), 5867–73.

128. Regl, G., Kasper, M., Schnidar, H., Eichberger, T., Neill, G. W., Ikram, M. S., Quinn, A. G., Philpott, M. P., Frischauf, A. M., and Aberger, F. (2004) The zinc-finger transcription factor Gli2 antagonizes contact inhibition and differentiation of human epidermal cells. *Oncogene*, **23**(6), 1263–74.

129. Regl, G., Kasper, M., Schnidar, H., Eichberger, T., Neill, G. W., Philpott, M. P., Esterbauer, H., Hauser-Kronberger, C., Frischauf, A. M., and Aberger, F. (2004) Activation of the BCL2 promoter in response to Hedgehog/Gli signal transduction is predominantly mediated by Gli2. *Cancer Res*, **64**(21), 7724–31.

130. Hosoya, T., Arai, M. A., Koyano, T., Kowithayakorn, T., and Ishibashi, M. (2008) Naturally occurring small-molecule inhibitors of Hedgehog/Gli-mediated transcription. *Chembiochem*, **9**(7), 1082–92.

131. Mimeault, M., Johansson, S. L., Vankatraman, G., Moore, E., Henichart, J. P., Depreux, P., Lin, M. F., and Batra, S. K. (2007) Combined targeting of epidermal growth factor receptor and Hedgehog signaling by gefitinib and cyclopamine cooperatively improves the cytotoxic effects of docetaxel on metastatic prostate cancer cells. *Mol Cancer Ther*, **6**(3), 967–78.

8 Targeting the Notch signaling pathway in cancer stem cells

Joon T. Park, Ie-Ming Shih, and Tian-Li Wang

Johns Hopkins Medical Institutions

NOTCH SIGNALING IN CANCER STEM CELLS

Stem cells are characterized by two unique properties: self-renewal and multilineage differentiation potential. Self-renewal provides the cell with the ability to go through numerous cycles of cell division, while maintaining a stem cell population through asymmetric cell division. For each division, a stem cell divides into two cells: another stem cell and a progenitor cell. It is thought that the stem cell retains the stem cell characteristics, while the progenitor cell can differentiate into tissue-specific cells within a limited number of cell divisions. Embryonic stem cells are active during embryonic differentiation and develop into all of the tissues in the body. Adult stem cells can be found in differentiated tissues and can differentiate into the entirety of cell types in the tissue from which they originate. Normal stem cells are transformed into cancer stem cells by acquiring somatic mutations in oncogenes or tumor suppressor genes.[1] Cancer stem cells share stem cell properties with embryonic stem cells such as self-renewal and differentiation potential.[2] Evidence suggests that many cancers, including leukemia, breast cancer, and glioma, contain a rare population of cells that are highly tumorigenic, in contrast to the bulk of cancer cells that have a limited capacity to form tumors in vivo. Cancer stem cells proliferate slowly, have indefinite self-replication ability, and are highly resistant to chemotherapy. Although conventional chemotherapy may eradicate the majority of cancer cells, cancer stem cells are largely spared and may go on to accumulate additional somatic mutations, eventually giving rise to recurrences and metastases. On the basis of this cancer stem cell theory, therapeutic strategies that specifically target pathways for cell renewal and cell fate decision in cancer stem cells could potentially increase the efficacy of current cancer treatment and reduce the risk of relapse and metastasis. Cancer stem cells isolated as a dye-excluding side population from numerous cancer cell lines express high levels of Notch receptors.[3,4] Consequently, Notch signaling is considered one of the most attractive targets for developing therapeutics directed at cancer stem cells.

NOTCH SIGNALING PATHWAY

The Notch signaling pathway is evolutionarily highly conserved and mediates intercellular signaling.[5] Notch signaling controls cell fate decision and patterned differentiation in numerous developmental processes. The Notch transmembrane receptors are activated by cell surface ligands DSL (Delta/Serrate/Lag2) and mediate direct cell–cell communication. Notch receptors are large, single-pass, type I transmembrane proteins. Four members, Notch1, 2, 3, and 4, have been identified in mammals. Notch receptor is synthesized as a single precursor protein but is cleaved by a furin-like protease at a juxtamembrane site (S1 cleavage) within the Golgi apparatus to create a heterodimer. Notch heterodimer consists of noncovalently associated extracellular and transmembrane subunits located at the plasma membrane.[6–8] The extracellular subunit contains a variable number of EGF-like repeats that are critical for ligand binding[9,10] and a juxtamembrane negative regulatory region (NRR) consisting of three LIN12/Notch repeats (LNR1–3) and a heterodimerization domain (HD1 or HD-N). The transmembrane subunit contains a small stretch of extracellular heterodimerization domain (HD2 or HD-C), a single-pass transmembrane domain, and an intracellular region that contains ankyrin repeats[11] involved in signal transduction.[12] In addition, this intracellular region contains nuclear localization signals (NLS) and a transactivation domain (TAD), followed by a PEST region rich in Proline, Glutamate, Serine, and Threonine residues that is involved in protein degradation (Figure 8–1).

In mammals, five Notch ligands have been identified: two Jagged (JAG1 and JAG2) and three Delta-like (Dll1, Dll3, and Dll4). All five ligands share a conserved

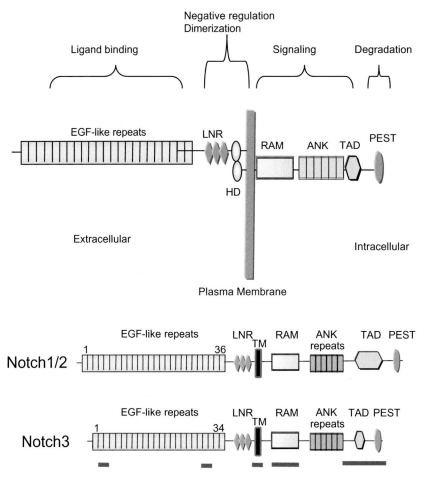

Figure 8–1: Diagram of Notch receptor. All Notch receptors consist of two subunits on the cell surface. RAM, CSL-interacting domain; ANK, ankyrin repeats; TAD, trans-activating domain; Pro-Glu-Ser-Thr (PEST), degradation motif. *See color plates.*

N-terminal DSL motif essential for binding to Notch receptor. Binding of DSL ligand to the N-terminal EGF-repeat region of the Notch extracellular subunit initiates a conformational change in the Notch receptor and triggers two sequential cleavages within the transmembrane subunit that are catalyzed by tumor necrosis factor-α converting enzyme (TACE) (S2 cleavage) at the extracellular surface, and by γ-secretase (S3 cleavage) at the intramembrane region. The latter cleavage releases Notch intracellular cytoplasmic domain (NICD) from the plasma membrane. NICD then shuttles into the nucleus, where it interacts with DNA-binding factors CSL (CBF1/Suppressor of Hairless /Lag1). In the absence of NICD, CSL acts as a repressor through interaction with corepressors and histone deacetylase.[13] Interaction of NICD and CSL disrupts the repressor complex and recruits coactivators, including mastermind-like 1 (MAML1) and histone acetyltransferase, which act in concert to activate Notch target gene expression (Figure 8–2).

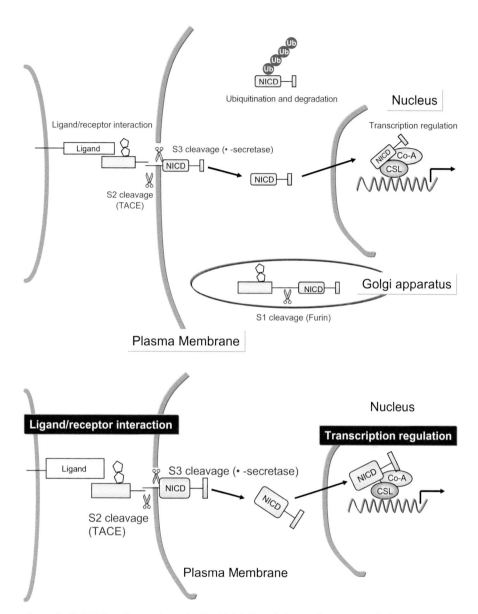

Figure 8–2: Notch pathway elements. The Notch ligands have a large extracellular region composed of a DSL domain and a series of EGF-like repeats. The Notch receptor is expressed on the plasma membrane as a heterodimer, which is generated during its translocation from the cytoplasm to the membrane by cleavage of the precursor protein by the convertase furin, and is glycosylated by Fringe. Upon ligand binding, Notch undergoes two consecutive cleavage steps: one by the protease TACE and one by a γ-secretase that results in the release of NICD and its nuclear translocation. NICD interacts with the transcription factor CSL, dissociating CSL from corepressor molecules (CoR) and recruiting coactivators (CoA) such as mastermind and histone acetyl transferase, giving rise to a transcription activation complex. Potential steps for therapeutic targeting of Notch signaling include receptor/ligand interaction, S1–S3 cleavages, transcription regulation, and protein degradation. *See color plates.*

Recent studies have demonstrated that the NRR maintains the "off" state of Notch receptors prior to ligand-induced activation. In human T-ALL, somatic mutations were frequently found in HD1 and HD2 domains of the NRR region. These mutations caused ligand-independent S2 cleavage and subsequent S3 cleavage, ultimately releasing NICD. Somatic mutations in the PEST domain that caused a premature stop, and thus resulted in deletion of the PEST domain, were also identified in T-ALL. Loss of the PEST domain results in increased NICD levels due to decreased proteasomal degradation.

NOTCH SIGNALING AS A CANCER THERAPEUTIC TARGET

Notch signaling plays an important role in cell fate decision and cell renewal. Deregulated expression of Notch receptors, ligands, and target genes has been found in a wide variety of hematological malignancies and solid tumors, including breast, ovarian, cervical, lung, head and neck, pancreas, and medulloblastoma. Therefore Notch inhibition may represent a promising treatment for cancer. In addition to the oncogenic potential, the Dll/Notch pathway is clearly required for cardiovascular development and is involved in angiogenesis. For example, genetic deletions of several Notch ligands, receptors, and downstream target genes cause vascular defects. Of note, Notch ligands, especially Dll4, were found to be strongly expressed in tumor endothelial cells, and inhibition of Dll4 has been shown to be a potential antitumor therapeutic strategy.

The main pharmacological strategies for targeting components of the Notch signaling pathway include blockage of receptor/ligand interactions and inhibition of the release of NICD. For example, (1) small compounds employed as inhibitors have been shown to prevent S1, S2, or S3 enzymatic cleavages, which are essential steps for activation of Notch signaling; (2) recombinant ligand proteins have been developed that act as competitive inhibitors for receptor binding; and (3) monoclonal antibodies have been developed to block activation of Notch receptor signaling. Potential strategies for future Notch-based targeted therapy include suppressing the protein levels of Notch components as well as inhibiting CSL/Notch transcription (Figure 8–2). In the following, we will briefly discuss the therapeutic perspectives of each strategy.

Blocking proteolytic cleavage events

Because the activation of Notch signaling involves three proteolytic cleavages of Notch receptor, blocking the enzyme responsible for each cleavage event may suppress the generation of NICD and thus block Notch signaling.

Blockade of S1 cleavage

Furin belongs to the family of pro-protein convertases that process precursor proteins into biologically active products.[14] Furin cleaves full-length Notch in the trans-Golgi network, which generates the mature heterodimeric cell surface receptor.[6,7] An inhibitor of furin proteases, AT-EK1, was found to inhibit the generation of mature heterodimer.[15] AT-EK1 treatment was also found to decrease the level of CSL-luciferase reporter activity in a dose-dependent manner, indicating that the formation of a mature heterodimer is required for Notch signaling.[15]

Blockade of S2 cleavage

TACE is a protease in ADAM (A disintegrin and metalloprotease) protease family that is involved in the S2 cleavage of Notch receptor. It is thought that ligand binding to Notch receptor induces TACE cleavages, which remove the extracellular domain of the transmembrane subunit of Notch receptor. TACE inhibitors have been developed previously. For example, BB3103 was found to inhibit ligand-induced activation of a CSL-responsive-reporter construct.[16]

Blockade of S3 cleavage

The most well-known small compound inhibitors of the Notch signaling pathway belong to the γ-secretase inhibitors (GSI), which block the S3 cleavage and inhibit the release of the intracellular domain of Notch.[17] Gamma-secretase is a large, integral membrane protease complex that includes a catalytic subunit and accessory subunits. The gamma-secretase complex catalyzes the intramembranous proteolysis of several membrane proteins, among which the Notch receptors and the beta-amyloid precursor peptide are of paramount therapeutic relevance. Promising results have been shown in in vitro and in vivo models. For example, in APC mutant mice, GSI treatment reduced expression of the Notch target gene Hes1 and converted proliferative adenoma cells into mucus-secreting goblet cells.[18] In a Kaposi's sarcoma (KS) mouse model, intratumoral injection of GSI inhibited tumor growth via decreasing proliferation and increasing apoptosis.[19] Accumulating preclinical evidence has led to the opening of phase I/phase II dose escalation clinical trials of a GSI in relapsed or refractory T-ALL and breast cancer patients. GSI is relatively easy and inexpensive to produce. However, GSI treatment also causes widespread side effects, including severe diarrhea.[20] The major drawback of GSI is its nonspecificity, as it targets many membrane proteins, including Notch receptors and ligands, Erb4, syndecan, and CD44. Although decreasing the GSI dose can ameliorate diarrhea, GSI treatment is always associated with chronic side effects. Thus short-term treatment is preferable, and benefits versus side effects should be evaluated carefully for each patient in future clinical trials.

Blocking receptor/ligand interaction

Recombinant Notch extracellular domain protein

The Notch extracellular domain is composed of 29–36 EGF-like repeats, and a recombinant protein containing as few as two of these modules is able to interact with Notch ligands and acts as a competitive inhibitor for binding to the full-length Notch.[21] Treatment of adipocytes with recombinant protein containing EGF-like repeats 11 and 12 has been shown to block their differentiation,[21] as expected based on the known role of Notch signaling in cell fate determination of mammalian cells.

Neutralizing DLL ligands

A soluble form of the extracellular domain of Dll4 (D4ECD) was found to efficiently block the Dll4/Notch pathway. Interestingly, D4ECD increased vascular density, but the blood vessels were poorly functional, thereby accounting for the delay in tumor growth.[22] When D4ECD was overexpressed in tumor cells, it served as a soluble inhibitor of Dll4/Notch signaling.[22] When these transgenic tumor cells were implanted into mice, the secreted soluble D4ECD reduced Notch signaling in the host and affected vascular morphologies as well as tumor growth rates. Increased vascular density with dramatic vessel sprouting and branching, which led to induced hypoxia and necrosis in the tumor, was observed in the D4ECD-expressing group.[22,23]

Neutralizing anti-Dll4 antibody

Neutralizing antibodies that bind to the Notch ligand or Notch receptor have been shown to be effective in inhibiting Notch signaling. A Dll4-neutralizing antibody was found to block the Dll4/Notch pathway.[24] Pro-angiogenesis and antitumor phenotypes were observed in tumors treated with the neutralizing anti-Dll4 antibody, similar to the Dll4-neutralizing ligand D4ECD mentioned in the previous section. Vascular targeting therapy using anti–Vascular endothelial growth factor (VEGF) can become ineffective, possibly as a consequence of the ability of tumors to adapt and become resistant to the treatment. On the other hand, Dll4-neutralizing antibody treatment can block the growth of tumors that are resistant to anti-VEGF.[24] Therefore Dll4-neutralizing antibody treatment provides an alternative option for delaying tumor growth when used in combination with anti-VEGF treatment.

Anti-Notch antibody

Monoclonal antibodies that inhibit proteolytic cleavages in Notch3 receptor have recently been developed.[25] Epitope mapping reveals that those inhibitory antibodies bind to two domains in the NRR, including LNR1 and HD2, suggesting that they clamp LNR1 and HD2 together to suppress the S2 cleavage. This finding supports the view that the NRR is important for autoinhibitory regulation.[26] Interestingly, stimulatory antibodies were also identified, and they were found to bind to linear epitope in LNR1 or HD2 of the NRR.[25] Therefore it would be interesting to analyze the conformational alteration that reveals the S2 cleavage site upon ligand stimulation and to develop strategies to either block or stimulate this process. It can be envisioned that in the future, antibodies that can modulate Notch activity are likely to have significant experimental and therapeutic potential.

Suppressing Notch protein expression

Ubiquitination controls Notch protein expression

Several distinct classes of E3 ubiquitin ligases appear to directly regulate the protein levels of Notch receptor. In mammals, the E3 ubiquitin ligase Itch has been shown to ubiquitinate membrane-tethered Notch1.[27] Furthermore, Sel10, an F-box/WD40 repeat-containing protein that interacts with an SCF ubiquitin ligase complex, also functions as a negative regulator in the Notch signaling pathway[28] by minimizing the half-life of NICD through ubiquitination.[29,30] Genetic study of *Drosophila* has demonstrated that Numb is a negative regulator of the Notch signaling pathway.[31] The study suggested that Numb recruits components of the ubiquitination machinery to the Notch receptor, thereby facilitating Notch1 ubiquitination at the membrane, which in turn promotes degradation of the Notch intracellular domain. Immunohistochemical analysis showed that Numb expression was lost in about 50% of human breast cancers[32] and that Numb silencing increased Notch signaling in Numb-positive breast tumors.[32] Understanding the negative regulation of Notch signaling seems likely to facilitate the development of future Notch-based novel therapies.

RNA interference against components of Notch signaling

Because of promising results obtained from in vitro studies, RNA interference (RNAi) against components of the Notch signaling pathway, including both receptor and ligand, is thought to hold promise as a future novel therapy.[33–35] The advantage of the RNAi strategy is that the molecules can be designed easily and the cost is relatively low. However, clinical use of RNAi therapy is likely a long way off because there are several challenges to overcome. These include how to achieve efficient delivery of RNAi into the appropriate tissue and how to increase the stability of RNAi. In addition, there is the possibility that excess RNAi will cause untoward immune response in the host.[36,37] Therefore there is a lot to be learned on the road to developing RNAi therapy.

Inhibiting the action of NICD

NICD is responsible for triggering transcription of Notch downstream genes by forming the NICD–CSL complex and recruiting coactivators such as MAML1. Therefore strategies to interrupt the formation of the NICD–CSL complex or to block the recruitment of coactivators can potentially suppress the expression of downstream target genes.

Dominant-negative MAML1

The MAML1 binding site on NICD is essential for the transcriptional activation of Notch signaling.[38,39] Truncated forms of MAML1 that retain affinity for NICD, but lack the activator domain, competitively inhibit the recruitment of wild-type MAML1 to the NICD–CSL complex. Blocking Notch signaling by this dominant negative MAML1 suppressed primary melanoma cell growth both in vitro and in vivo.[40] Dominant-negative MAML1 peptide (62-amino-acid peptide) formed a transcriptionally inactive nuclear complex with NICD and CSL and inhibited the growth of both murine and human Notch1-transformed T-ALLs.[41] Therefore blocking the formation of a functional NICD coactivator complex can be a potential strategy for chemotherapeutic intervention.

CONCLUSION

In the past few years, evidence has accumulated that supports the cancer stem cell theory, which suggests that targeting cancer stem cells may provide a new strategy to treat human cancer. However, developing strategies to selectively inactivate or eradicate cancer stem cells appears to be challenging because Notch signaling pathways are shared by both cancer stem cells and normal stem cells. A variety of recent studies using antagonists that target these pathways suggests that it may be feasible to selectively target the cancer stem cell population. For example, Notch signaling is required for the differentiation and proliferation of cancer stem cells but is not required for the maintenance of blood-forming stem cells.[42] Exciting new findings concerning the involvement of Notch signaling in cancer stem cells have placed this signal transduction pathway in the focus of therapeutic target development in cancer treatment. However, our understanding of Notch signaling in specific cell types and diseases still remains unclear. Very little is known about the functional relationship between each Notch ligand and its Notch receptor in specific tissue microenvironments, or about the different downstream targets of each Notch receptor. Because many conundrums regarding the role of Notch signaling in both normal and cancer stem cells remain, Notch signaling and the development of strategies for the clinical targeting of the Notch pathway deserve further study.

ACKNOWLEDGMENTS

This work was supported by a New Investigator Research Grant from DoD Ovarian Cancer Research Program (T.-L.W.), a Research Scholar Grant from the American Cancer Society (T.-L.W.), and a career development award from NIH cervical cancer SPORE (T.-L.W.).

REFERENCES

1. Reya, T., et al., *Stem cells, cancer, and cancer stem cells.* Nature, 2001. **414**(6859): pp. 105–111.
2. Huntly, B.J.P., and D.G. Gilliland, *Leukaemia stem cells and the evolution of cancer-stem-cell research.* Nat Rev Cancer, 2005. **5**(4): pp. 311–321.
3. Patrawala, L., et al., *Side population is enriched in tumorigenic, stem-like cancer cells, whereas ABCG2+ and ABCG2− cancer cells are similarly tumorigenic.* Cancer Res, 2005. **65**(14): pp. 6207–6219.
4. Fan, X., et al., *Notch pathway inhibition depletes stem-like cells and blocks engraftment in embryonal brain tumors.* Cancer Res, 2006. **66**(15): pp. 7445–7452.
5. Bray, S.J., *Notch signalling: a simple pathway becomes complex.* Nat Rev Mol Cell Biol, 2006. **7**(9): pp. 678–689.
6. Blaumueller, C.M., et al., *Intracellular cleavage of Notch leads to a heterodimeric receptor on the plasma membrane.* Cell, 1997. **90**(2): pp. 281–291.
7. Logeat, F., et al., *The Notch1 receptor is cleaved constitutively by a furin-like convertase.* Proc Natl Acad Sci U S A, 1998. **95**(14): pp. 8108–8112.

8. Brou, C., et al., *A novel proteolytic cleavage involved in Notch signaling: the role of the disintegrin-metalloprotease TACE.* Mol Cell, 2000. **5**(2): pp. 207–216.

9. Fehon, R.G., et al., *Molecular interactions between the protein products of the neurogenic loci Notch and Delta, two EGF-homologous genes in Drosophila.* Cell, 1990. **61**(3): pp. 523–534.

10. Rebay, I., et al., *Specific EGF repeats of Notch mediate interactions with Delta and serrate: implications for Notch as a multifunctional receptor.* Cell, 1991. **67**(4): pp. 687–699.

11. Tamura, K., et al., *Physical interaction between a novel domain of the receptor Notch and the transcription factor RBP-J[kappa]/Su(H).* Curr Biol, 1995. **5**(12): pp. 1416–1423.

12. Blank, V., et al., *NF-[kappa]B and related proteins: Rel/dorsal homologies meet ankyrin-like repeats.* Trends Biochem Sci, 1992. **17**(4): pp. 135–140.

13. Tun, T., et al., *Recognition sequence of a highly conserved DNA binding protein RBP-Jx.* Nucl Acids Res, 1994. **22**(6): pp. 965–971.

14. Nakayama, K., *Furin: a mammalian subtilisin/Kex2p-like endoprotease involved in processing of a wide variety of precursor proteins.* Biochem J, 1997. **327**(Pt 3): pp. 625–635.

15. Bush, G., et al., *Ligand-induced signaling in the absence of furin processing of Notch1.* Dev Biol, 2001. **229**(2): pp. 494–502.

16. Moloney, D.J., et al., *Fringe is a glycosyltransferase that modifies Notch.* Nature, 2000. **406**(6794): pp. 369–375.

17. LaVoie, M.J., and D.J. Selkoe, *The Notch ligands, Jagged and Delta, are sequentially processed by [alpha]-secretase and presenilin/[gamma]-secretase and release signaling fragments.* J Biol Chem, 2003. **278**(36): pp. 34427–34437.

18. van Es, J.H., et al., *Notch/[gamma]-secretase inhibition turns proliferative cells in intestinal crypts and adenomas into goblet cells.* Nature, 2005. **435**(7044): pp. 959–963.

19. Curry, C.L., et al., *Gamma secretase inhibitor blocks Notch activation and induces apoptosis in Kaposi's sarcoma tumor cells.* Oncogene, 2005. **24**(42): pp. 6333–6344.

20. Barten, D.M., et al., *Gamma-secretase inhibitors for Alzheimer's disease: balancing efficacy and toxicity.* Drugs R D, 2006. **7**(2): pp. 87–97.

21. Garces, C., et al., *Notch-1 controls the expression of fatty acid-activated transcription factors and is required for adipogenesis.* J Biol Chem, 1997. **272**(47): pp. 29729–29734.

22. Noguera-Troise, I., et al., *Blockade of Dll4 inhibits tumour growth by promoting nonproductive angiogenesis.* Nature, 2006. **444**(7122): pp. 1032–1037.

23. Scehnet, J.S., et al., *Inhibition of Dll4-mediated signaling induces proliferation of immature vessels and results in poor tissue perfusion.* Blood, 2007. **109**(11): pp. 4753–4760.

24. Ridgway, J., et al., *Inhibition of Dll4 signalling inhibits tumour growth by deregulating angiogenesis.* Nature, 2006. **444**(7122): pp. 1083–1087.

25. Li, K., et al., *Modulation of Notch signaling by antibodies specific for the extracellular negative regulatory region of NOTCH3.* J Biol Chem, 2008. **283**(12): pp. 8046–8054.

26. Gordon, W.R., et al., *Structural basis for autoinhibition of Notch.* Nat Struct Mol Biol, 2007. **14**(4): pp. 295–300.

27. Qiu, L., et al., *Recognition and ubiquitination of Notch by Itch, a Hect-type E3 ubiquitin ligase.* J Biol Chem, 2000. **275**(46): pp. 35734–35737.

28. Hubbard, E.J.A., et al., *Sel-10, a negative regulator of lin-12 activity in* Caenorhabditis elegans, *encodes a member of the CDC4 family of proteins.* Genes Dev, 1997. **11**(23): pp. 3182–3193.

29. Gupta-Rossi, N., et al., *Functional interaction between SEL-10, an F-box protein, and the nuclear form of activated Notch1 receptor.* J Biol Chem, 2001. **276**(37): pp. 34371–34378.

30. Oberg, C., et al., *The Notch intracellular domain is ubiquitinated and negatively regulated by the mammalian Sel-10 homolog.* J Biol Chem, 2001. **276**(38): pp. 35847–35853.

31. McGill, M.A., and C.J. McGlade, *Mammalian numb proteins promote Notch1 receptor ubiquitination and degradation of the Notch1 intracellular domain.* J Biol Chem, 2003. **278**(25): pp. 23196–23203.

32. Pece, S., et al., *Loss of negative regulation by Numb over Notch is relevant to human breast carcinogenesis.* J Cell Biol, 2004. **167**(2): pp. 215–221.

33. Park, J.T., et al., *Notch3 gene amplification in ovarian cancer.* Cancer Res, 2006. **66**(12): pp. 6312–6318.

34. Wang, Z., et al., *Down-regulation of Notch-1 contributes to cell growth inhibition and apoptosis in pancreatic cancer cells.* Mol Cancer Ther, 2006. **5**(3): pp. 483–493.

35. Purow, B.W., et al., *Expression of Notch-1 and its ligands, Delta-Like-1 and Jagged-1, is critical for glioma cell survival and proliferation.* Cancer Res, 2005. **65**(6): pp. 2353–2363.

36. Sledz, C.A., et al., *Activation of the interferon system by short-interfering RNAs.* Nat Cell Biol, 2003. **5**(9): pp. 834–839.

37. Rossi, J., et al., *Wandering eye for RNAi.* Nat Med, 2008. **14**(6): p. 611.

38. Wu, L., et al., *MAML1, a human homologue of* Drosophila *Mastermind, is a transcriptional co-activator for NOTCH receptors.* Nat Genet, 2000. **26**(4): pp. 484–489.

39. Kitagawa, M., et al., *A human protein with sequence similarity to* Drosophila *Mastermind coordinates the nuclear form of Notch and a CSL protein to build a transcriptional activator complex on target promoters.* Mol Cell Biol, 2001. **21**(13): pp. 4337–4346.

40. Liu, Z.-J., et al., *Notch1 signaling promotes primary melanoma progression by activating mitogen-activated protein kinase/phosphatidylinositol 3-kinase-Akt pathways and up-regulating N-cadherin expression.* Cancer Res, 2006. **66**(8): pp. 4182–4190.

41. Weng, A.P., et al., *Growth suppression of pre-T acute lymphoblastic leukemia cells by inhibition of Notch signaling.* Mol Cell Biol, 2003. **23**(2): pp. 655–664.

42. Maillard, I., et al., *Canonical Notch signaling is dispensable for the maintenance of adult hematopoietic stem cells.* Cell Stem Cell, 2008. **2**(4): pp. 356–366.

9 TGF-β, Notch, and Wnt in normal and malignant stem cells: differentiating agents and epigenetic modulation

Stephen Byers, Michael Pishvaian, Lopa Mishra, and Robert Glazer
Lombardi Comprehensive Cancer Center, Georgetown University

The notion that the growth and so-called aberrant differentiation of many tumors depend on the existence of a small population of cancer stem cells in much the same way that organogenesis and tissue replacement depend on normal stem cells is at the heart of contemporary investigations of neoplastic diseases. Not surprisingly, the same genetic and signaling pathways that are involved in normal stem cell renewal and specification are also important in tumorigenesis. These pathways include the Wnt/β-catenin, Notch, and TGF-β signaling systems, all of which have been reviewed recently.[1-5] In this chapter, we will highlight areas that are either developing or have not been covered extensively in other reviews. For example, recent studies have highlighted a role for the RNA binding protein Musashi 1 (Msi1) in the regulation of normal and cancer stem cells through the Wnt and Notch pathways.[6,7]

Notch and Wnt signaling also regulate and are regulated by asymmetric cell division, a defining stem cell characteristic that has received little attention in the cancer stem cell literature.[8-10] Asymmetric cell division, which results in the segregation of damaged proteins into only one of the daughter cells, has also recently been linked to stem cell aging, a process that clearly differs between normal and cancer stem cells. The ability of carcinoma cells to take on characteristics typical of cells from quite different backgrounds is well established and almost certainly related to a pluripotent stem cell–like origin. Perhaps the best example of this is the ability of carcinoma cells to acquire the molecular and phenotypic hallmarks of migratory and invasive mesenchymal cells.[11,12] Melanomas and several other tumor cells can also take on the genotypic and phenotypic characteristics of blood vessels, called *vascular mimicry*, which can also be induced by differentiating agents such as vitamin A.[13–16] This latter observation indicates that, although such agents might inhibit proliferation of a subpopulation of tumor cells, they might stimulate the transdifferentiation of another subpopulation (cancer stem cells). These data raise the interesting question of why particular lineage-determining genes, but not others, have the potential to become activated during tumor progression. A recent study demonstrates that differential histone methylation identifies a set of lineage-specific genes that are transcriptionally poised (but not expressed) in multipotential hematopoietic stem/progenitor cells, which have the potential to become activated during the process of terminal differentiation.[17] These data broaden the scope of our discussion to include differentiating agents and epigenetic regulation of cancer stem cell activity.

TGF-β FAMILY

The TGF-β family contains about 30 structurally related growth and differentiation factors that include TGF-βs, activins, Nodal, and bone morphogenic proteins (BMPs). TGF-β family signals are conveyed through two types (type I and type II) of transmembrane receptor serine–threonine kinases, which together form a tetrameric receptor complex at the cell surface.[18] Ligand binding to this complex induces a conformational change that induces phosphorylation and activation of type I by type II receptors. Activation of Smad transcription factors results in their nuclear translocation and subsequent Smad-mediated activation or repression of gene expression. The Smad activation and activity are modulated by a variety of receptor- or Smad-interacting proteins that include ubiquitin and small ubiquitin-like modifier ligases as well as multiple proteins in the transcription complexes. Depending on the differentiation stage of the target cell, the local environment, and the identity and dosage of the ligand, TGF-β proteins promote or inhibit cell proliferation, apoptosis, and differentiation. The diverse and often seemingly contradictory TGF-β functions can be understood by gene dosage, cross-talk of TGF-β family signaling through Smads with other signaling pathways such as Wnt and Hedgehog signaling and receptor tyrosine kinase signaling, and interactions of Smads with a multitude of DNA binding transcription factors, which themselves are targeted by signaling pathways.[18,19]

TGF-β family signaling in embryonic stem (ES) cells

Although human and mouse ES cells show substantial differences in their requirements, TGF-β family proteins play a role in both the maintenance of the cells in their undifferentiated state and in the selection and initiation of differentiation. Nodal or activin, two TGF-β–related proteins that share the same receptors and Smads, are thought to naturally play a role in human ES cell maintenance because their receptors are expressed and their Smads are activated in undifferentiated ES cells. The activated Smad pathway downstream from these ligands is thus likely to cooperate with Wnt signaling in keeping the ES cells undifferentiated and pluripotent. BMP signals, in cooperation with lymphocyte inhibitory factor, a member of the IL-6 cytokine family, have also been shown to be required for keeping mouse ES cells undifferentiated. Apparently, BMPs maintain the ES cells undifferentiated through activation of BMP Smads, which, in turn, activate Id transcription factors, a small class of basic helix-loop-helix (bHLH) transcription factors that act as inhibitors of differentiation. *Id* proteins, in turn, insulate the cell from lineage priming by complementing gp130/STAT3 activity.

TGF-β family signals mediate key decisions that specify germ layer differentiation. Thus activin induces ventral or dorsal mesoderm and endoderm in *Xenopus* explants, depending on the dosage. In mammals, this signaling pathway is presumably activated by Nodal or related factors that activate the same pathway as activin, thus giving rise to mesoderm and endoderm. Conversely, inhibition of both activin/TGF-β and BMP signaling gives rise to neuroectoderm formation in *Xenopus*, while the absence of these factors also allows neuroectoderm formation from mouse ES cells in culture. Activin or TGF-β also induces mesoderm differentiation, whereas BMP signals confer ectodermal and mesodermal differentiation of human ES cells. Activin signaling also leads ES cells to differentiate into endoderm. Thus TGF-β family signaling or the absence of TGF-β family signaling is a determinant of both maintenance and initial specification of ES cells and of the primary cell fate decision in early embryogenesis that will give rise to multiple cell lineages and cell fates.

Several recent reports suggest functional interactions between TGF-β/Smad and Notch signaling in various tissues, based on hierarchical activation of one pathway by the other or by coordinate regulation of common target genes. A subset of the family of classical Notch target genes, bHLH transcriptional repressors of the hairy/enhancer-of-split-related (H/Espl) family, including *HEY1*, *HEY2*, *HES1*, and *HES5*, and the Notch ligand *JAGGED1* are induced by TGF-β at the onset of epithelial–mesenchymal transformation (EMT) in a panel of epithelial cells from mammary gland, kidney tubules, and epidermis.[20] TGF-β–induced EMT is prevented by silencing of HEY1 or Jagged1 and by chemical inactivation of Notch. These findings suggest the functional integration of TGF-β/Smad3 and Jagged1/Notch signaling in EMT.[20] Interestingly, LEF1, the downstream target of WNT/β-catenin signaling, is activated by TGF-β3 in a β-catenin–independent, Smad-dependent process, during EMT underlying the process of palatal fusion.[21] Ectopic expression of LEF1 in the presence of stabilized nuclear β-catenin can also induce EMT directly[22]; however, the role of TGF-β signaling in this process, if anything, remains unclear.

TGF-β family members in neural stem cells

Neural differentiation from uncommitted ES cells is thought to occur in the absence of exogenous TGF-β family factors, yet is also regulated by other inhibitory factors and cell adhesion proteins.[23,24] Thus BMPs inhibit neural differentiation and promote epidermal differentiation in *Xenopus* embryo explants and inhibit neural differentiation of ES cells, yet a gradient of BMP signaling defines the dorsoventral patterning of the neural tube.[25] BMPs inhibit embryonic day 13 ventricular zone progenitor cells but enhance astroglial and neural crest cell differentiation at day 16, at higher doses, inducing apoptosis.[26–28] Intriguingly, BMP-2 suppresses Sonic hedgehog (Shh)-induced proliferation of medulloblastoma granule precursor cells, displaying a tumor suppressive role. Additionally, loss of function of Nodal or cripto, a functional Nodal receptor complex, also enhances neurogenesis, whereas addition of Nodal suppresses neural differentiation.[29,30] Thus TGF-β family signals, irrespective of which Smad pathway is activated, act to prevent initiation of neural lineage selection.[31,32] Later in development, TGF-β promotes differentiation and lineage expansion of established progenitors, for example, by inducing autonomic gangliogenesis or olfactory neuron proliferation.[33] The intricate orchestration of neuronal development by various components of the TGF-β pathway is exemplified in the decrease of cerebellar Purkinje cells in Smad4$^{-/-}$ mice and a proliferation of precursor cells in the developing cortex of mice lacking the Smad adaptor protein ELF.[34,35] Recent evidence suggests that once a precursor lineage is established, TGF-β signaling accelerates the differentiation and lineage commitment of precursor cells that can accumulate in its absence. Once cells are fully differentiated, TGF-β inhibits growth of glial cells, setting the stage for malignant transformation into gliomas, when the growth inhibitory response to TGF-β is somehow inactivated. Thus, when the delicate control of normal development by TGF-β family factors is disrupted, tumors may ensue. In gliomas, TGF-β stimulates tumor progression and invasiveness, concomitantly with the angiogenic and immunomodulating activities of increased TGF-β expression by these cells.[36,37]

TGF-β signaling in hematopoietic stem cells

TGF-β family proteins and their downstream signaling effectors, the Smads, also play key roles in hematopoietic differentiation.[38,39] TGF-β itself inhibits the proliferation of early multipotent hematopoietic stem cells, but not of later progenitors. The effects of TGF-β on more mature progenitor cells are complex and depend on the presence of other growth factors.[40] In contrast to TGF-β, BMPs, in combination with cytokines, promote hematopoietic specification, differentiation, and proliferation of human ES cells.[41] While TGF-β acts as a negative regulator of hematopoietic progenitor and stem cells in vitro, impaired TGF-β signaling in vivo does not affect hematopoietic lineage selection.[40] Indeed, the absence of a functional type I TGF-β receptor allows for normal development of hematopoietic progenitors and functional hematopoiesis in mouse embryos, even though they die at midgestation, with severe defects in vascular development of the yolk sac and placenta and an absence of circulating red blood cells.[42] Conversely, the absence of Smad5, an effector of BMP signaling and, as more recently appreciated, of TGF-β signaling, enhances the efficiency of hematopoietic progenitor cell generation in embryoid bodies derived from ES cells,[43] further supporting the notion that TGF-β accelerates the differentiation and proliferation of committed precursors. Signaling by TGF-β family proteins through Smads also regulates cell fate commitment decisions of myeloid versus lymphoid precursors, as illustrated by enhanced myeloid differentiation at the expense of lymphoid commitment, when Smad7, which inhibits Smad activation, is overexpressed.[44] To add to the complexity, the activation of feedback loops by TGF-β family signaling further defines the regulation of hematopoietic stem cell differentiation. For example, BMPs activate the expression of the homeobox transcription factor Dlx1, which, in turn, blocks activin-induced differentiation of a hematopoietic cell line by interacting with Smad4 through its homeodomain.[45]

TGF-β signals in gastrointestinal tissues and cancers

Nowhere is the role of TGF-β family signaling at the interface between development and cancer more prominent than in gut epithelial cells.[46–49] Several TGF-β signaling components are bona fide tumor suppressors, with the ability to constrain cell growth and inhibit cancer development at its early stages. Inactivation of at least one of these components (such as the TGF-β receptors Smad2 or Smad4) occurs in almost all gastrointestinal tumors.[48,49] In the colon, stem cells are found at the bases of crypts between the villi. With continuing cell differentiation from this stem cell region, often referred to as a *niche*, the differentiating cells move upward to the tip of the villi, where they undergo apoptosis.[58] TGF-β signaling mediators, such as the type II TGF-β receptor Smad4 and the Smad adaptor ELF, are all expressed in the crypts, suggesting an active role for TGF-β signaling in the stem cell compartment. As cells acquire a malignant phenotype and cancer progresses from adenoma to metastatic cancer, expression of these proteins is successively lost, indicating their role in tumor suppression. Inactivation of at least one of these components, such as the type II TGF-β receptor Smad2 or Smad4, occurs in almost all gastrointestinal tumors.[51] Mouse experiments further illustrate the role of TGF-β signaling effectors in the suppression of gastrointestinal carcinoma development. For example, *Smad4*[+/−] mice develop gastric tumors, and intercrossing of the *Smad4*[+/−] genotype into mice with a mutation in the adenomatous polyposis coli tumor suppressor APC[Δ716] results in the development of larger and more invasive colorectal tumors than in the presence of the two *Smad4* alleles.[52] These findings are consistent with the role of Smad4 in normal gut endoderm development. Defects in gastrointestinal epithelial cell shape and polarity are also seen in *Smad2*[+/−] and *Smad3*[+/−] double heterozygous and *elf*[−/−] homozygous mice, further arguing for a role of TGF-β signaling in normal gastrointestinal epithelial development.[53] Interestingly, many aspects of the *elfβ*-spectrin phenotype are alike to those of the *Drosophila* β-spectrin,[34] which is, in turn, reminiscent of the *labial* phenotype, particularly in the gut. Control of the homeotic gene *labial* is dependent on activation by extracellular gradients of *wingless* and *decapentaplegic* (the *Drosphila* homologue of TGF-β) during embryogenesis.[34,54]

BMP signaling may also play an active role in the stem cell compartments of the colon, presumably by suppressing the effects of Wnt signaling and, consequently, limiting stem cell renewal. Consistent with this notion, mutations in the BMP receptor BMPR1A and in Smad4 contribute to juvenile intestinal polyposis and Cowden disease, respectively. Furthermore, inactivation of the gene for one of the type I BMP receptors in mice allows for an expansion of the stem and progenitor cell populations, eventually leading to intestinal polyposis resembling human juvenile polyposis syndrome.[42] Finally, TGF-β signaling also appears to be important for the transition of stem cells to a progenitor and fully differentiated phenotype in the liver and biliary system. Accordingly, $Smad2^{+/-}$ and $Smad3^{+/-}$ double heterozygous and $elf^{-/-}$ homozygous mice all show defective liver development, with $elf^{+/-}$ mice developing hepatocellular carcinoma.[55] Moreover, the TGF-β– and BMP-regulated protein PRAJA is expressed in hepatoblasts and modulates ELF and Smad3[54,56] (and STKE pathway CMP_17699). The absence of this drive to normal epithelial differentiation may thus favor formation of human hepatocellular carcinoma. TGF-β family proteins and their signaling pathways play key roles in the self-renewal and maintenance of stem cells in their undifferentiated state, while changes in TGF-β family signals drive the selection of defined differentiation pathways and their progression of differentiation. When deregulated, changes in TGF-β family signaling may contribute to impaired differentiation and allow for the development of cancers, thus linking the differentiation of stem cells with suppression of carcinogenesis.

WNT SIGNALING

Wnt signaling is essential for maintenance of stem cell compartments in several organs and regulates cellular differentiation.[2,3,57] This effect of Wnt signaling can be mimicked by stabilizing cytoplasmic/nuclear β-catenin alone, and, as discussed earlier, the effects of Wnt signaling on stem cells are modulated through association with other signaling pathways, including Notch, Shh, and TGF-β signaling. Understanding of the regulation of this stem cell compartment has come mostly from the examination of intestinal, epidermal, and hematopoietic cells.

The intestine provides an excellent model for studying stem cell proliferation and differentiation. The details of the intestinal cellular architecture are covered by Radtke and Clevers.[58] Intestinal stem cells reside in the base of colonic crypts (or at position +4 in the small intestine) and give rise to transient amplifying and ultimately differentiated cells such that each intestinal crypt is monoclonal in origin. Crypt cells rise toward the lumen as they differentiate and are eventually shed by apoptosis at the luminal surface. In the intestine, Wnt signaling maintains the crypt stem cell compartment, and Wnt signaling induces proliferation of the transient amplifying cells. The essential nature of Wnt signaling is demonstrated by the fact that in $tcf4^{-/-}$ mice, the intestinal proliferative compartment is entirely absent.[59] Furthermore, transgenic overexpression of the Wnt antagonist Dickkopf-1 results in the loss of intestinal crypts in adult mice.[60] Activation of Wnt signaling is present in 90% of all sporadic human colorectal cancers, though not through overactivity of Wnt itself; rather, 70% to 80% of tumors contain inactivating mutations in APC, while an additional 10% to 20% have degradation-resistant mutations in β-catenin.[61] These mutations result in the sustained expression of Wnt target genes such as c-Myc and cyclin D1.

Maintenance of hematopoietic stem cells (HSC) is also dependent on Wnt function.[62–65] The stem cells themselves secrete Wnt ligand to help maintain the stem cell microenvironment. These effects can be mimicked through activation of the Wnt signaling pathway by a degradation-resistant β-catenin, while HSC compartment reconstitution can be inhibited by overexpression of Axin.[66] Wnt signaling is dysregulated in hematologic malignancies such as chronic lymphocytic leukemia, pre-B acute lymphocytic leukemia, and the blast crises of chronic myelogenous leukemia.[67–69] Surprisingly, the prognostic implications of overexpression of Wnt-regulated genes have not been as well explored in hematologic malignancies as they have in solid tumors.

Wnt/β-catenin activation is also highly oncogenic in the mammary gland. Transgenic mice expressing either constitutively active MMTV-ΔN89 β-catenin or MMTV-Wnt1 show increased lobuloalveolar development and precocious tumor growth.[70,71] These transgenic models exhibit marked increases in Sca-1⁺/CK6⁺ stem/progenitor cells[72] and side population cells,[73] and isografts of Wnt1-transduced mammary epithelial cells give rise to both luminal epithelial and myoepithelial cells.[74] β-Catenin coactivation is also involved in the maintenance of embryonic stem cell pluripotency,[75] and a similar mechanism may pertain to other cell populations such as mesenchymal stem cells.[76] GSK3β inhibitors that activate the Wnt pathway are able to maintain mouse and human embryonic stem cell pluripotency.[77–79]

The Wnt pathway is inhibited by members of the Dickkopf (DKK) gene family,[80] secreted proteins that comprise three species. DKK1 and DKK2, but not DKK3, directly interfere with Wnt binding to the LRP-5/6 coreceptor[81] and function as tumor suppressors. DKK3 (also known as Reduced Expression in Immortalized Cells, or REIC) may serve a similar function since its expression is reduced in many cancers.[82] DKK3 prevents nuclear localization of β-catenin,[83] and its expression in prostate cancer cells disrupts acinar morphogenesis and growth, although not by inhibition of β-catenin/T-cell factor (TCF) activity.[84] Reduced DKK3 expression in melanoma cells results in loss of cell adhesion, increased invasion, upregulation of the transcriptional repressor Snail-1,[85] and reduction of E-cadherin,[86] characteristics indicative of EMT. Musashi (Msi1)-mediated reduction of DKK3 increased β-catenin stability, its nuclear localization, and activation of both Wnt and Notch signaling,[7] and reduction of DKK3 by shRNA in mammary epithelial cells reverts them to the Msi1 phenotype.[7] Overall, the literature suggests that Msi1 complements Notch and Wnt signaling in the context of mammary stem cell expansion and transformation.

NOTCH AND Msi

In the epidermis, mammary gland, and gut, interplay among the Wnt, Hedgehog, BMP, and Notch pathways determines whether stem cells self-renew or differentiate. For example, Wnt signaling is activated in the colonic crypt and maintains cells in a proliferative state; increased activity of the Wnt pathway leads to enlarged crypts and intestinal tumors, whereas Wnt inhibition results in loss of the stem cell compartment altogether. Notch acts jointly with Wnt to sustain stem cell proliferation and is essential for the differentiation of specific cell types. Hedgehog signaling promotes differentiation and restricts crypt formation that is mediated through its effect on BMP signaling.[87] The Notch pathway also plays an important role in stem cell self-renewal and cell fate determination, particularly in breast stem cells.[88] Notch promotes both stem cell self-renewal and differentiation in a context-dependent manner,[89] and thus other signaling pathways are likely to influence whether Notch functions as a tumor suppressor or an oncogene in a particular tissue. Notch is activated by sequential proteolytic cleavage of its membrane-associated form to a constitutively active intracellular form (NIC)[90] (Figure 9–1). Notch activation is influenced by at least two factors: the negative regulator Numb[91] and the positive regulator Msi1.[92] Msi1 and Notch are markers of hematopoietic, neuroglial, hair follicle, intestinal, testis, and breast stem cells,[93–97] suggesting a global role in cell fate determination. Numb promotes the ubiquitination and proteasomal degradation of Notch[98] and interferes with nuclear translocation by binding to its C-terminal proline, glutamic acid, serine, threonine (PEST) sequence.[91] Tissues that strongly express Msi1 show almost no expression of Numb,[99] suggesting that Msi may be a critical modulator of Notch activity. Although Msi1 is presumed to block translation in the cytoplasm, it is found in high abundance in the nucleus of *Drosophila* cells[100] and intestinal crypt cells,[101] suggesting that Msi1 may also interfere with pre-mRNA splicing, processing, and/or nuclear-cytoplasmic transport. In mice with increased IGF2 signaling due to loss of maternal imprinting,[101] Msi1 is highly expressed in the nucleus of intestinal crypt cells and correlates with an increase in stem cells and a predisposition to tumorigenesis. Msi1 expression is impacted by a second family of RNA-binding proteins related to *Drosophila* Elav, a protein involved in the development and maintenance of the nervous system in the fly and mouse.[102,103] The mammalian Elav orthologs, HuB, HuC, and HuD, promote mRNA stabilization by binding to AU-rich elements in the 3′-UTRof Msi1 and other mRNAs,[104] and its mRNA stabilization activity is linked to protein kinase C α activity.[105] It is interesting that HuB, HuC, and HuD are expressed in neuronal stem cells in a manner similar to Msi1[106] and are similarly localized to the nucleus.[107] Although Msi1 was originally identified as an inhibitor of Numb

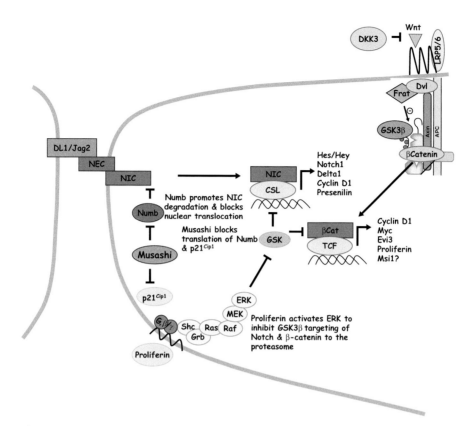

Figure 9–1: Msi signaling pathways associated with mammary stem cells. Notch is processed proteolytically to an extracellular domain (NEC) and an intracellular domain (NIC). The Notch ligands Delta1 (DL1) and Jagged2 (Jag2) associate with NEC and induce cleavage and release of membrane-bound NIC. NIC translocates to the nucleus, where it serves as a coactivator of CSL to activate transcription of Hes/Hey, Notch, DL1/Jag2, and cyclin D1. Msi inhibits the translation of Numb and p21 (cip1) by binding to a motif in the 3-UTR. Inhibition of Numb prevents NIC degradation and nuclear translocation, whereas inhibition of p21 (cip1) prevents inhibition of cyclin-dependent kinases to promote G1/S transition. Msi increases secretion of the growth factor proliferin, which induces ERK activation through the Gi-coupled IGFII receptor, resulting in inhibition of GSK3β. Msi1 blocks expression of the Wnt pathway inhibitor, DKK3, to increase β-catenin/T-cell factor (TCF) activity. Msi1 is a TCF target gene and is transcriptionally activated by c-Myc, also a TCF target gene. *See color plates.*

translation,[108] it also blocks translation of *Drosophila* Tramtrack69 (Ttk69),[109] a transcriptional repressor that negatively regulates EGFR expression in insect cells.[110] In addition, Msi1 blocks translation of p21[Cip1] (Figure 9–1), an inhibitor of cyclin-dependent protein kinases that regulate transit through G1/S of the cell cycle[111] as well as chromosome segregation and mitosis.[112] These effects may account for increased proliferation secondary to reduction of p21[Cip1] in Msi1-expressing cells.[7] Reduction of p21[Cip1] by Msi1 may also account for activation of β-catenin/TCF-dependent transcription[7] since p21[Cip1] negatively regulates Wnt4 transcription.[113] Because β-catenin participates in establishing the mitotic spindle,[114] this effect may also contribute to the enhanced proliferative activity elicited by Msi1.[7] Interestingly, the Msi1 promoter contains several TCF-binding elements,[115] suggesting that Msi1 itself may be a target gene and autoregulated through the Wnt pathway. The activated intracellular form of Notch1 as well as the Notch ligands Jagged and Delta are highly expressed in breast cancer[116–118] and in Msi1-expressing cells,[7] in contrast to the little or no expression of Numb.[119] Transgenic mouse models have documented the involvement of Notch signaling in mammary tumorigenesis. Mammary-directed expression of the intracellular activated form of Notch1, Notch3, or Notch4 results in a spectrum of tumors, ranging from hyperplasia to poorly differentiated carcinomas.[120,121] Notch3 and Notch4 increase lobuloalveolar development from luminal epithelial cells, and transgene expression of all Notch subtypes results in lactation failure. Notch-dependent transformation is associated with ERK activation downstream of the Ras pathway,[122] which increases Notch mRNA stability[123] and is required for transcription of the Notch target gene Hes-1.[124] Similar to β-catenin/TCF signaling, c-Myc and cyclin D1 are Notch target genes in mouse and human mammary tumors[125] and in MMTV-Notch1 transgenic mice.[126]

Notch3 gene expression is increased in human breast spheroid cultures (mammospheres), which are enriched in stem and progenitor cells.[127] Stimulation of Notch-dependent transcription increases colony formation, whereas inhibition of Notch4 blocks branching in Matrigel,[50] characteristics that are similar to the phenotype of Notch4 transgenic mice.[128] Intracellular Notch1 is increased in breast epithelium enriched in Msi1, label-retaining cells, and CK19+/ER+ side population cells of the large light cell morphology.[129] Msi1-expressing cells were ER+ and CK14/CK18 double-positive and consisted of luminal and myoepithelial progenitor cells with branch-forming capacity in Matrigel,[130] characteristics similar to those reported by Dontu and colleagues in breast stem cells[50] and in Msi1-expressing mammary epithelial cells.[7]

ASYMMETRIC CELL DIVISION AND STEM CELL AGING

According to Gonczy,

> Asymmetric division occurs when a mother cell gives rise to two daughter cells with different fates. Sometimes, the two daughters are identical at birth, and the fate difference is established later on, for instance through signaling from neighboring cells. Alternatively, the mother cell can be polarized and the two daughters are distinct already at birth. A sequence of four steps can be recognized in most instances of such intrinsic asymmetric cell divisions. First, symmetry is broken in the mother cell. Second, the mother cell becomes polarized. Third, fate determinants are segregated towards given regions of the polarized mother cell. Fourth, the mitotic spindle is aligned such that cleavage results in the correct partitioning of determinants to the daughter cells, thus ensuring that the daughters have different fates. As a result of these four steps, a mother cell can generate two daughter cells that are born at the same time but that are not identical.[9]

Although Notch and Wnt signaling also regulate and are regulated by asymmetric stem cell division, this area has not been addressed in any detail with respect to cancer stem cells. In *Caenorhabditis elegans*, asymmetric division is controlled by the asymmetric activity of a Wnt signaling pathway (the Wnt/β-catenin asymmetry pathway). In this process, two specialized β-catenin homologs have crucial roles in the transmission of Wnt signals to the asymmetric activity of a TCF-type transcription factor, POP-1, in the daughter cells. One β-catenin homolog regulates the distinct nuclear level of POP-1, and the other functions as a coactivator of POP-1. Both β-catenins localize asymmetrically in the daughter nuclei using different mechanisms.[131] In *Drosophila* neurogenesis, unequal distribution of the Notch inhibitor Numb during mitosis inhibits Notch activity in one daughter to induce neuronal differentiation.[9] In contrast, vertebrate Numb accumulates at adherens junctions in the apical end feet, which anchor adjacent progenitor cells. Numb and the Numb-related protein Numblike, which is present throughout the cytoplasm, have a partially redundant role during early neurogenesis, and their combined absence results in depletion of progenitor cells and overproduction of neurons. This phenotype reflects a requirement for cadherin-based adhesion, as adherens junctions are disrupted in animals that lack Numb and Numblike, and as an analogous phenotype is observed after cadherins have been depleted from wild-type cells. Therefore, although Numb and Numblike are important for asymmetric cell division in the vertebrate cerebral cortex, they function differently than *D. melanogaster* Numb. The Numb inhibitor Msi1 also functions during asymmetric cell division to control neuroglial differentiation.[100]

One of the first essential events to drive asymmetric cell division is mediated by a complex of proteins that includes the polarity protein Par6. Interactions between Par6 and the TGF-β pathway could potentially yield crucial new information in the generation of cancer stem/progenitor cells.[132] For instance, asymmetric localization of Par6 along with the Bazooka/Par3 and PKCα pro-differentiation factors controls cell fate.[133] The Par6 complex, in turn, regulates the activity of the tumor suppressor Lgl (lethal giant larval); Lgl then localizes PKC to the *Drosophila* neuroblast cortex and daughter self-renewing cells. Importantly, Lgl is involved in the secretion of Decapentaplegic (Dpp), a member of the TGF-β family, as well as in the expression of Dpp targets tinman, labial, and, consequently, dorsoventral patterning of the *Drosophila* ectoderm. Interestingly, the type I TGF-β receptor has been found to interact with the Par6 N-terminal PB1 domain at tight junctions of epithelial cells. Thus it is possible that polarity and patterning are also controlled by the TGF-β family in conjunction with Par6/Lgl. The formation of polarized cells and apical junctions, such as tight junctions, must be a dynamic process, in which cell types, tissues, and organs are constantly becoming molded and dissolved as the organism grows. So it is not surprising that, in mouse mammary epithelial cells, tight junction dissolution by TGF-β occurs when the type II TGF-β receptor kinase associates with and phosphorylates Par6.[132] Par6 then recruits Smurf1, which then ubiquitinates RhoA, a small GTPase family member responsible for the maintenance of apicobasal polarity and junctional stability, resulting in the breakdown of tight junctions and EMT, a fundamental process in development and tumor progression.[134] These observations indicate the bifunctional role of TGF-β in potentially maintaining as well as dissolving asymmetry in a context-driven manner.

Although disturbances in the regulation of asymmetric cell division can result in unrestrained proliferation and tumors in flies, to our knowledge, no studies have examined this process in bona fide cancer stem cells. Presumably, cancer stem cells must engage in asymmetric cell division to result in self-renewal and differentiation to nonstem tumor cells. What, then, are the fundamental differences between cancer stem cells and normal stem cells? Both exhibit the ability to maintain telomere length, as do nonstem tumor cells, and both exhibit a degree of pluripotency. However, normal stem cells, even though they maintain their telomeres, do, in fact, age and senesce, whereas cancer stem cells either do not or do so at a lower rate and can undergo self-renewal even in the face of aging-associated damage. In keeping with their role in the maintenance of stem cell numbers, both the Notch and Wnt pathways are involved in the premature aging disease progeria.[135] Other studies have demonstrated impairment in Wnt signaling during the regulation and fate of stem cells in the muscle,[136] skin, and intestine.[137] Specific mutations in the human gene encoding lamin A or in the lamin A processing enzyme Zmpste24, which result in premature aging, affect adult stem cells by interfering with the Notch and Wnt signaling pathways.[135] These studies also imply that the same deterioration in stem cell regulation also occurs in normal aging. Indeed, defective lamin A is present in healthy individuals at low levels and accumulates as visible aggregates in old age.[138,139] Consistent with this, levels of undamaged lamin A are significantly reduced in aged hematopoietic stem cells. These data are examples of another important concept in cellular aging, that is, accumulation of damaged misfolded proteins leads to aging and a shorter life span.[135,140,141] How do these observations relate to asymmetric cell division and cancer stem cells? Remarkably, bacteria and yeast as well as vertebrates asymmetrically partition damaged proteins into one of the daughter cells following mitosis or budding.[142] Several studies have indicated that this also occurs during stem cell and asymmetric division, and even during somatic mitoses.[135] Although there is some debate regarding the nature of the cell that inherits the damaged proteins (mother or daughter), the cell that does inherit is usually the one with the shorter life span. Clearly this may differ depending on whether the organ is rapidly or slowly turning over.

In human embryonic stem cells and other mammalian cultured cells, segregation of proteins in the TGF-β and Wnt pathways destined for degradation (Smad1 phosphorylated by MAPK and GSK3, phospho-β-catenin, and total polyubiquitinylated proteins) is asymmetric.[143] Recent studies have emphasized the role of endosomal sorting complexes required for transport (ESCRT) in asymmetric cell division, autophagy, and cancer.[144–146] For example, in cells lacking the ESCRT component TSG101, a known tumor suppressor, Notch is trapped in the endosome, leading to enhanced signaling and cell-autonomous proliferation.[147] Interestingly, a number of ESCRT components are recruited to the midbody, a structure that is often inherited asymmetrically in dividing stem cells.[145,148,149]

THE ENVIRONMENT, DIFFERENTIATING AGENTS, AND CANCER STEM CELLS

Although the role of the environment and diet in the regulation of tumor initiation and progression is now well established, few studies have addressed their influence on cancer stem cells. The dietary agents vitamin A and vitamin D (VD) have anticancer properties as well as differentiation-promoting effects. For example, the active form of vitamin A, retinoic acid (RA), is important in embryogenesis and skin biology, and VD is important for bone health. For many years, both the differentiating and growth-inhibitory properties of RA and VD have been thought of as essential for their anticancer activity. We suggest that while this may be partly true when these agents are used to prevent the development of cancer, it may be less so after the cancer has begun and during its progression. In fact, we anticipate that differentiation during carcinogenesis, in which a population of cells may well harbor a stem or progenitor cell phenotype, is just as likely to result in a negative outcome (e.g., formation of blood vessel cells or mesenchymal [invasive] cells) as a positive one (e.g., normal breast or white blood cells). For example, some breast cancer cells transdifferentiate into cells resembling those that form blood vessels when treated with RA.[13] In other words, the differentiation-promoting activities of RA and VD may be responsible both for their failure as a treatment in certain circumstances and their side effects, a result of their action on nontumor cells.

The ability of carcinoma cells to take on characteristics typical of cells from quite different backgrounds is well established and may be related to a stem cell–like origin. Perhaps the best example of this is the ability of breast carcinoma cells to acquire the molecular and phenotypic hallmarks of migratory and invasive mesenchymal cells.[11,12] Melanomas and several other tumor cells can also take on the genotypic and phenotypic characteristics of blood vessels, known as vascular mimicry.[14–16] As discussed previously, our data suggest that these properties need to be taken into account when considering treatment or prevention regimens using potent differentiating agents such as vitamins A and D. Although some agents might inhibit proliferation of the majority of tumor cells (nonstem cancer cells), they might stimulate the transdifferentiation of another population (cancer stem cells). In more advanced cancers, the antiproliferative effects of differentiation agents, such as RA and VD, are often lost. However, it is not known if the small proportion of self-renewing cancer stem cells in these tumors can still differentiate in response to these agents. If they can, then one can easily imagine a scenario in which RA treatment could promote transdifferentiation of the cancer stem cells in these otherwise nonresponsive tumors. Differentiation to a mesenchymal or vascular lineage obviously would not be a desired response. Clearly those tumor cells that reacquire differentiated properties in response to such an agent still harbor the genetic defects that transformed them initially and *cannot* be considered as normal differentiated cells. RA and VD can inhibit the activity of the Wnt pathway so important for the maintenance of stem cell number.[150–153] In the case of vitamin A, induction of the HMG protein and potential β-catenin inhibitor Sox9 was associated with its ability to promote endothelial transdifferentiation of breast cancer cells.[13] This is consistent with the observation that initial stem cell specification in the skin critically depends on Sox9.[154] In this study, all three epidermal epithelial lineages were lost in the absence of Sox9. In the case of breast cancer (stem?) cells treated with vitamin A, we hypothesized that induction of Sox9 resulted in cells being inappropriately directed down an endothelial, rather than an epithelial, lineage.[13] In this latter example, dozens of endothelium-specific genes that were not expressed in the absence of treatment were induced by vitamin A. This raises the question, why were genes specifying this particular lineage induced, and not genes specifying another lineage? A clue may be provided by recent experiments in which methylation/demethylation of lineage-specifying gene promoters and/or covalent histone modifications was shown to identify lineage-specific genes that were transcriptionally poised (but not expressed) in stem/progenitor cells.[17,52,155]

CAN EPIGENETIC MODIFICATIONS IDENTIFY POISED LINEAGE-PREDICTING GENES IN CANCER STEM CELLS?

The fundamental basis of cell specification during embryogenesis and organogenesis is conversion of cell fate decisions into heritable epigenetic information that determines cellular identity. Methylation/demethylation of lineage-specifying gene promoters and covalent histone modifications are heritable epigenetic markers that play a central role in this process. In ES cells, inactive genes encoding early developmental regulators possess bivalent histone modification domains and are therefore poised for activation. During erythroid development, H3K4me2 and H3K4me3 are concordant at most genes; however, multipotential hematopoietic cells have a subset of genes that are differentially methylated (H3K4me2+/me3−).[17] These genes are transcriptionally silent, highly enriched in lineage-specific hematopoietic genes, and uniquely susceptible to differentiation-induced H3K4 demethylation. Self-renewing ES cells, which restrict H3K4 methylation to genes containing CpG islands (CGIs), lack H3K4me2+/me3−-containing genes. These data reveal distinct epigenetic regulation of CGI and non-CGI genes during development and indicate an interactive relationship between DNA sequence and differential H3K4 methylation in lineage-specific differentiation. In this case, incomplete methylation of H3K4 identifies a set of lineage-specific genes that are transcriptionally poised in multipotential hematopoietic stem/progenitor cells and that have the potential to become activated during the process of terminal differentiation. However, bivalent histone domains are not always observed at typical tissue-specific genes. For example, windows of unmethylated CpG dinucleotides and putative pioneer factor interactions, such as members of the FOXA family, mark enhancers for at least some tissue-specific genes in ES cells.[52,155] The unmethylated windows expand in cells that express the gene and contract, disappear, or remain unchanged in nonexpressing tissues. However, in ES cells, they do not always coincide with common histone modifications. These findings indicate that pioneer factor interactions in ES cells promote the assembly of a chromatin structure that is permissive for subsequent activation and in which differentiated tissues lack the machinery required for gene activation when these ES cell markers are absent. These enhancer markers, like histone modifications, may therefore represent important features of the pluripotent state. Taken together, these studies provide a view of epigenetic changes that occur during development that reveal a complex interdependence among DNA sequence, histone modifications, and developmental gene function. Identification of poised lineage or tissue-specific genes before they are actually transcribed in pluripotent cells during development may also provide insight into the fate of cancer cells. As indicated previously, carcinoma cells often acquire characteristics typical of cells from quite different backgrounds. Is it possible that the epigenetic marks discussed previously may allow us to identify tumor cells destined to undergo, or that are at least capable of, EMT or vascular transdifferentiation, for example, before they actually do so? In this way, we may be able to discriminate between morphologically indistinguishable cells or between tumors that have the potential to metastasize or vascularize and those that do not.

REFERENCES

1. Kitisin, K., Saha, T., Blake, T., Golestaneh, N., Deng, M., Kim, C., Tang, Y., Shetty, K., Mishra, B., and Mishra, L. (2007). Tgf-β signaling in development. *Sci STKE 2007*, (399):cm1. Review.
2. Mishra, L., Shetty, K., Tang, Y., Stuart, A., and Byers, S.W. (2005). The role of TGF-β and Wnt signaling in gastrointestinal stem cells and cancer. Oncogene **24**, 5775–5789.
3. Fodde, R., and Brabletz, T. (2007). Wnt/beta-catenin signaling in cancer stemness and malignant behavior. Curr Opin Cell Biol **19**, 150–158.
4. Chiba, S. (2006). Notch signaling in stem cell systems. Stem Cells **24**, 2437–2447.
5. Watt, F.M., Estrach, S., and Ambler, C.A. (2008). Epidermal Notch signalling: differentiation, cancer and adhesion. Curr Opin Cell Biol **20**, 171–179.
6. Glazer, R.I., Wang, X.Y., Yuan, H., and Yin, Y. (2008). Musashi1: a stem cell marker no longer in search of a function. Cell Cycle **7**, 2635–2639.
7. Wang, X.Y., Yin, Y., Yuan, H., Sakamaki, T., Okano, H., and Glazer, R.I. (2008). Musashi1 modulates mammary progenitor cell expansion through proliferin-mediated activation of the Wnt and Notch pathways. Mol Cell Biol **28**, 3589–3599.
8. Green, J.L., Inoue, T., and Sternberg, P.W. (2008). Opposing Wnt pathways orient cell polarity during organogenesis. Cell **134**, 646–656.
9. Gonczy, P. (2008). Mechanisms of asymmetric cell division: flies and worms pave the way. Nat Rev Mol Cell Biol **9**, 355–366. Quote from p. 355.
10. Le Borgne, R., and Schweisguth, F. (2003). Unequal segregation of Neuralized biases Notch activation during asymmetric cell division. Dev Cell **5**, 139–148.
11. Birchmeier, C., Birchmeier, W., and Brand-Saberi, B. (1996). Epithelial-mesenchymal transitions in cancer progression. Acta Anat (Basel) **156**, 217–226.
12. Sommers, C.L., Thompson, E.W., Torri, J.A., Kemler, R., Gelmann, E.P., and Byers, S.W. (1991). Cell adhesion molecule uvomorulin expression in human breast cancer cell lines: relationship to morphology and invasive capacities. Cell Growth Differ **2**, 365–372.

13. Endo, Y., Deonauth, K., Prahalad, P., Hoxter, B., Zhu, Y., and Byers, S.W. (2008). Role of Sox-9, ER81 and VE-cadherin in retinoic acid-mediated trans-differentiation of breast cancer cells. PLoS ONE **3**, e2714.

14. Hendrix, M.J., Seftor, E.A., Kirschmann, D.A., and Seftor, R.E. (2000). Molecular biology of breast cancer metastasis: molecular expression of vascular markers by aggressive breast cancer cells. Breast Cancer Res **2**, 417–422.

15. Hendrix, M.J., Seftor, E.A., Hess, A.R., and Seftor, R.E. (2003). Vasculogenic mimicry and tumour-cell plasticity: lessons from melanoma. Nat Rev Cancer **3**, 411–421.

16. Maniotis, A.J., Folberg, R., Hess, A., Seftor, E.A., Gardner, L.M., Pe'er, J., Trent, J.M., Meltzer, P.S., and Hendrix, M.J. (1999). Vascular channel formation by human melanoma cells in vivo and in vitro: vasculogenic mimicry. Am J Pathol **155**, 739–752.

17. Orford, K., Kharchenko, P., Lai, W., Dao, M.C., Worhunsky, D.J., Ferro, A., Janzen, V., Park, P.J., and Scadden, D.T. (2008). Differential H3K4 methylation identifies developmentally poised hematopoietic genes. Dev Cell **14**, 798–809.

18. Attisano, L., and Wrana, J.L. (2002). Signal transduction by the TGF-β superfamily. Science **296**, 1646–1647.

19. Letamendia, A., Labbe, E., and Attisano, L. (2001). Transcriptional regulation by Smads: crosstalk between the TGF-β and Wnt pathways. J Bone Joint Surg Am **83A**(Suppl 1), 31–39.

20. Zavadil, J., Cermak, L., Soto-Nieves, N., and Bottinger, E.P. (2004). Integration of TGF-β/Smad and Jagged1/Notch signalling in epithelial-to-mesenchymal transition. EMBO J **23**, 1155–1165.

21. Nawshad, A., and Hay, E.D. (2003). TGFbeta3 signaling activates transcription of the LEF1 gene to induce epithelial mesenchymal transformation during mouse palate development. J Cell Biol **163**, 1291–1301.

22. Kim, K., Lu, Z., and Hay, E.D. (2002). Direct evidence for a role of beta-catenin/LEF-1 signaling pathway in induction of EMT. Cell Biol Int **26**, 463–476.

23. Munoz-Sanjuan, I., and Brivanlou, A.H. (2002). Neural induction, the default model and embryonic stem cells. Nat Rev Neurosci **3**, 271–280.

24. Temple, S. (2001). The development of neural stem cells. Nature **414**, 112–117.

25. Ying, Q.L., Nichols, J., Chambers, I., and Smith, A. (2003). BMP induction of Id proteins suppresses differentiation and sustains embryonic stem cell self-renewal in collaboration with STAT3. Cell **115**, 281–292.

26. Fogarty, M.P., Kessler, J.D., and Wechsler-Reya, R.J. (2005). Morphing into cancer: the role of developmental signaling pathways in brain tumor formation. J Neurobiol **64**, 458–475.

27. Graham, A., Francis-West, P., Brickell, P., and Lumsden, A. (1994). The signalling molecule BMP4 mediates apoptosis in the rhombencephalic neural crest. Nature **372**, 684–686.

28. Kleber, M., Lee, H.Y., Wurdak, H., Buchstaller, J., Riccomagno, M.M., Ittner, L.M., Suter, U., Epstein, D.J., and Sommer, L. (2005). Neural crest stem cell maintenance by combinatorial Wnt and BMP signaling. J Cell Biol **169**, 309–320.

29. Parisi, S., D'Andrea, D., Lago, C.T., Adamson, E.D., Persico, M.G., and Minchiotti, G. (2003). Nodal-dependent Cripto signaling promotes cardiomyogenesis and redirects the neural fate of embryonic stem cells. J Cell Biol **163**, 303–314.

30. Strizzi, L., Bianco, C., Normanno, N., and Salomon, D. (2005). Cripto-1: a multifunctional modulator during embryogenesis and oncogenesis. Oncogene **24**, 5731–5741.

31. Stern, C.D. (2005). Neural induction: old problem, new findings, yet more questions. Development **132**, 2007–2021.

32. Tropepe, V., Hitoshi, S., Sirard, C., Mak, T.W., Rossant, J., and van der Kooy, D. (2001). Direct neural fate specification from embryonic stem cells: a primitive mammalian neural stem cell stage acquired through a default mechanism. Neuron **30**, 65–78.

33. Gangemi, R.M., Perera, M., and Corte, G. (2004). Regulatory genes controlling cell fate choice in embryonic and adult neural stem cells. J Neurochem **89**, 286–306.

34. Tang, Y., Katuri, V., Dillner, A., Mishra, B., Deng, C.X., and Mishra, L. (2003). Disruption of transforming growth factor-beta signaling in ELF beta-spectrin-deficient mice. Science **299**, 574–577.

35. Zhou, Y.X., Zhao, M., Li, D., Shimazu, K., Sakata, K., Deng, C.X., and Lu, B. (2003). Cerebellar deficits and hyperactivity in mice lacking Smad4. J Biol Chem **278**, 42313–42320.

36. Jennings, M.T., and Pietenpol, J.A. (1998). The role of transforming growth factor beta in glioma progression. J Neurooncol **36**, 123–140.

37. Uhl, M., Aulwurm, S., Wischhusen, J., Weiler, M., Ma, J.Y., Almirez, R., Mangadu, R., Liu, Y.W., Platten, M., Herrlinger, U., Murphy, A., Wong, D.H., Wick, W., Higgins, L.S., and Weller, M. (2004). SD-208, a novel transforming growth factor beta receptor I kinase inhibitor, inhibits growth and invasiveness and enhances immunogenicity of murine and human glioma cells in vitro and in vivo. Cancer Res **64**, 7954–7961.

38. Ruscetti, F.W., Akel, S., and Bartelmez, S.H. (2005). Autocrine transforming growth factor-beta regulation of hematopoiesis: many outcomes that depend on the context. Oncogene **24**, 5751–5763.

39. Scandura, J.M., Boccuni, P., Massague, J., and Nimer, S.D. (2004). Transforming growth factor beta-induced cell cycle arrest of human hematopoietic cells requires p57KIP2 up-regulation. Proc Natl Acad Sci U S A **101**, 15231–15236.

40. Larsson, J., and Karlsson, S. (2005). The role of Smad signaling in hematopoiesis. Oncogene **24**, 5676–5692.

41. Park, C., Afrikanova, I., Chung, Y.S., Zhang, W.J., Arentson, E., Fong, G.G., Rosendahl, A., and Choi, K. (2004). A hierarchical order of factors in the generation of FLK1- and SCL-expressing hematopoietic and endothelial progenitors from embryonic stem cells. Development **131**, 2749–2762.

42. Larsson, J., Goumans, M.J., Sjostrand, L.J., van Rooijen, M.A., Ward, D., Leveen, P., Xu, X., Ten Dijke, P., Mummery, C.L., and Karlsson, S. (2001). Abnormal angiogenesis but intact hematopoietic potential in TGF-β type I receptor-deficient mice. EMBO J **20**, 1663–1673.

43. Liu, B., Sun, Y., Jiang, F., Zhang, S., Wu, Y., Lan, Y., Yang, X., and Mao, N. (2003). Disruption of Smad5 gene leads to enhanced proliferation of high-proliferative potential precursors during embryonic hematopoiesis. Blood **101**, 124–133.

44. Chadwick, K., Shojaei, F., Gallacher, L., and Bhatia, M. (2005). Smad7 alters cell fate decisions of human hematopoietic repopulating cells. Blood **105**, 1905–1915.

45. Chiba, S., Takeshita, K., Imai, Y., Kumano, K., Kurokawa, M., Masuda, S., Shimizu, K., Nakamura, S., Ruddle, F.H., and Hirai, H. (2003). Homeoprotein DLX-1 interacts with Smad4 and blocks a signaling pathway from activin A in hematopoietic cells. Proc Natl Acad Sci U S A **100**, 15577–15582.

46. Goumans, M.J., and Mummery, C. (2000). Functional analysis of the TGF-β receptor/Smad pathway through gene ablation in mice. Int J Dev Biol **44**, 253–265.

47. Kulkarni, A.B., and Karlsson, S. (1997). Inflammation and TGF-β 1: lessons from the TGF-β 1 null mouse. Res Immunol **148**, 453–456.

48. Massague, J., Blain, S.W., and Lo, R.S. (2000). TGF-β signaling in growth control, cancer, and heritable disorders. Cell **103**, 295–309.

49. Weinstein, M., Yang, X., and Deng, C. (2000). Functions of mammalian Smad genes as revealed by targeted gene disruption in mice. Cytokine Growth Factor Rev **11**, 49–58.

50. Dontu, G., Jackson, K.W., McNicholas, E., Kawamura, M.J., Abdallah, W.M., and Wicha, M.S. (2004). Role of Notch signaling in cell-fate determination of human mammary stem/progenitor cells. Breast Cancer Res **6**, R605–R615.

51. Takaku, K., Oshima, M., Miyoshi, H., Matsui, M., Seldin, M.F., and Taketo, M.M. (1998). Intestinal tumorigenesis in compound mutant mice of both Dpc4 (Smad4) and Apc genes. Cell **92**, 645–656.

52. Xu, J., Pope, S.D., Jazirehi, A.R., Attema, J.L., Papathanasiou, P., Watts, J.A., Zaret, K.S., Weissman, I.L., and Smale, S.T. (2007). Pioneer factor interactions and unmethylated

CpG dinucleotides mark silent tissue-specific enhancers in embryonic stem cells. Proc Natl Acad Sci U S A **104**, 12377–12382.

53. Tang, Y., Katuri, V., Srinivasan, R., Fogt, F., Redman, R., Anand, G., Said, A., Fishbein, T., Zasloff, M., Reddy, E.P., Mishra, B., and Mishra, L. (2005). Transforming growth factor-beta suppresses nonmetastatic colon cancer through Smad4 and adaptor protein ELF at an early stage of tumorigenesis. Cancer Res **65**, 4228–4237.

54. Monga, S.P., Tang, Y., Candotti, F., Rashid, A., Wildner, O., Mishra, B., Iqbal, S., and Mishra, L. (2001). Expansion of hepatic and hematopoietic stem cells utilizing mouse embryonic liver explants. Cell Transplant **10**, 81–89.

55. Bhowmick, N.A., Chytil, A., Plieth, D., Gorska, A.E., Dumont, N., Shappell, S., Washington, M.K., Neilson, E.G., and Moses, H.L. (2004). TGF-β signaling in fibroblasts modulates the oncogenic potential of adjacent epithelia. Science **303**, 848–851.

56. Mishra, L., Tully, R.E., Monga, S.P., Yu, P., Cai, T., Makalowski, W., Mezey, E., Pavan, W.J., and Mishra, B. (1997). Praja1, a novel gene encoding a RING-H2 motif in mouse development. Oncogene **15**, 2361–2368.

57. Nusse, R. (2008). Wnt signaling and stem cell control. Cell Res **18**, 523–527.

58. Radtke, F., and Clevers, H. (2005). Self-renewal and cancer of the gut: two sides of a coin. Science **307**, 1904–1909.

59. van de Wetering, M., Sancho, E., Verweij, C., de Lau, W., Oving, I., Hurlstone, A., van der Horn, K., Batlle, E., Coudreuse, D., Haramis, A.P., Tjon-Pon-Fong, M., Moerer, P., van den Born, M., Soete, G., Pals, S., Eilers, M., Medema, R., and Clevers, H. (2002). The beta-catenin/TCF-4 complex imposes a crypt progenitor phenotype on colorectal cancer cells. Cell **111**, 241–250.

60. Pinto, D., Gregorieff, A., Begthel, H., and Clevers, H. (2003). Canonical Wnt signals are essential for homeostasis of the intestinal epithelium. Genes Dev **17**, 1709–1713.

61. Morin, P.J., Sparks, A.B., Korinek, V., Barker, N., Clevers, H., Vogelstein, B., and Kinzler, K.W. (1997). Activation of beta-catenin-Tcf signaling in colon cancer by mutations in beta-catenin or APC. Science **275**, 1787–1790.

62. Congdon, K.L., and Reya, T. (2008). Divide and conquer: how asymmetric division shapes cell fate in the hematopoietic system. Curr Opin Immunol **20**, 302–307.

63. Reya, T., and Clevers, H. (2005). Wnt signalling in stem cells and cancer. Nature **434**, 843–850.

64. Duncan, A.W., Rattis, F.M., DiMascio, L.N., Congdon, K.L., Pazianos, G., Zhao, C., Yoon, K., Cook, J.M., Willert, K., Gaiano, N., and Reya, T. (2005). Integration of Notch and Wnt signaling in hematopoietic stem cell maintenance. Nat Immunol **6**, 314–322.

65. Rattis, F.M., Voermans, C., and Reya, T. (2004). Wnt signaling in the stem cell niche. Curr Opin Hematol **11**, 88–94.

66. Reya, T., Duncan, A.W., Ailles, L., Domen, J., Scherer, D.C., Willert, K., Hintz, L., Nusse, R., and Weissman, I.L. (2003). A role for Wnt signalling in self-renewal of haematopoietic stem cells. Nature **423**, 409–414.

67. Lu, D., Zhao, Y., Tawatao, R., Cottam, H.B., Sen, M., Leoni, L.M., Kipps, T.J., Corr, M., and Carson, D.A. (2004). Activation of the Wnt signaling pathway in chronic lymphocytic leukemia. Proc Natl Acad Sci U S A **101**, 3118–3123.

68. McWhirter, J.R., Neuteboom, S.T., Wancewicz, E.V., Monia, B.P., Downing, J.R., and Murre, C. (1999). Oncogenic homeodomain transcription factor E2A-Pbx1 activates a novel WNT gene in pre-B acute lymphoblastoid leukemia. Proc Natl Acad Sci U S A **96**, 11464–11469.

69. Jamieson, C.H., Ailles, L.E., Dylla, S.J., Muijtjens, M., Jones, C., Zehnder, J.L., Gotlib, J., Li, K., Manz, M.G., Keating, A., Sawyers, C.L., and Weissman, I.L. (2004). Granulocyte-macrophage progenitors as candidate leukemic stem cells in blast-crisis CML. N Engl J Med **351**, 657–667.

70. Imbert, A., Eelkema, R., Jordan, S., Feiner, H., and Cowin, P. (2001). Delta N89 beta-catenin induces precocious development, differentiation, and neoplasia in mammary gland. J Cell Biol **153**, 555–568.

71. Tsukamoto, A.S., Grosschedl, R., Guzman, R.C., Parslow, T., and Varmus, H.E. (1988). Expression of the int-1 gene in transgenic mice is associated with mammary gland hyperplasia and adenocarcinomas in male and female mice. Cell **55**, 619–625.

72. Li, Y., Welm, B., Podsypanina, K., Huang, S., Chamorro, M., Zhang, X., Rowlands, T., Egeblad, M., Cowin, P., Werb, Z., Tan, L.K., Rosen, J.M., and Varmus, H.E. (2003). Evidence that transgenes encoding components of the Wnt signaling pathway preferentially induce mammary cancers from progenitor cells. Proc Natl Acad Sci U S A **100**, 15853–15858.

73. Liu, B.Y., McDermott, S.P., Khwaja, S.S., and Alexander, C.M. (2004). The transforming activity of Wnt effectors correlates with their ability to induce the accumulation of mammary progenitor cells. Proc Natl Acad Sci U S A **101**, 4158–4163.

74. Naylor, S., Smalley, M.J., Robertson, D., Gusterson, B.A., Edwards, P.A., and Dale, T.C. (2000). Retroviral expression of Wnt-1 and Wnt-7b produces different effects in mouse mammary epithelium. J Cell Sci **113**(Pt 12), 2129–2138.

75. Miyabayashi, T., Teo, J.L., Yamamoto, M., McMillan, M., Nguyen, C., and Kahn, M. (2007). Wnt/beta-catenin/CBP signaling maintains long-term murine embryonic stem cell pluripotency. Proc Natl Acad Sci U S A **104**, 5668–5673.

76. Etheridge, S.L., Spencer, G.J., Heath, D.J., and Genever, P.G. (2004). Expression profiling and functional analysis of Wnt signaling mechanisms in mesenchymal stem cells. Stem Cells **22**, 849–860.

77. Trowbridge, J.J., Xenocostas, A., Moon, R.T., and Bhatia, M. (2006). Glycogen synthase kinase-3 is an in vivo regulator of hematopoietic stem cell repopulation. Nat Med **12**, 89–98.

78. Ding, S., Wu, T.Y., Brinker, A., Peters, E.C., Hur, W., Gray, N.S., and Schultz, P.G. (2003). Synthetic small molecules that control stem cell fate. Proc Natl Acad Sci U S A **100**, 7632–7637.

79. Sato, N., Meijer, L., Skaltsounis, L., Greengard, P., and Brivanlou, A.H. (2004). Maintenance of pluripotency in human and mouse embryonic stem cells through activation of Wnt signaling by a pharmacological GSK-3-specific inhibitor. Nat Med **10**, 55–63.

80. Krupnik, V.E., Sharp, J.D., Jiang, C., Robison, K., Chickering, T.W., Amaravadi, L., Brown, D.E., Guyot, D., Mays, G., Leiby, K., Chang, B., Duong, T., Goodearl, A.D., Gearing, D.P., Sokol, S.Y., and McCarthy, S.A. (1999). Functional and structural diversity of the human Dickkopf gene family. Gene **238**, 301–313.

81. Mao, B., Wu, W., Li, Y., Hoppe, D., Stannek, P., Glinka, A., and Niehrs, C. (2001). LDL-receptor-related protein 6 is a receptor for Dickkopf proteins. Nature **411**, 321–325.

82. Niehrs, C. (2006). Function and biological roles of the Dickkopf family of Wnt modulators. Oncogene **25**, 7469–7481.

83. Hoang, B.H., Kubo, T., Healey, J.H., Yang, R., Nathan, S.S., Kolb, E.A., Mazza, B., Meyers, P.A., and Gorlick, R. (2004). Dickkopf 3 inhibits invasion and motility of Saos-2 osteosarcoma cells by modulating the Wnt-beta-catenin pathway. Cancer Res **64**, 2734–2739.

84. Kawano, Y., Kitaoka, M., Hamada, Y., Walker, M.M., Waxman, J., and Kypta, R.M. (2006). Regulation of prostate cell growth and morphogenesis by Dickkopf-3. Oncogene **25**, 6528–6537.

85. Poser, I., Dominguez, D., de Herreros, A.G., Varnai, A., Buettner, R., and Bosserhoff, A.K. (2001). Loss of E-cadherin expression in melanoma cells involves up-regulation of the transcriptional repressor Snail. J Biol Chem **276**, 24661–24666.

86. Kuphal, S., Lodermeyer, S., Bataille, F., Schuierer, M., Hoang, B.H., and Bosserhoff, A.K. (2006). Expression of Dickkopf genes is strongly reduced in malignant melanoma. Oncogene **25**, 5027–5036.

87. Pitsouli, C., and Perrimon, N. (2008). Developmental biology: our fly cousins' gut. Nature **454**, 592–593.

88. Liu, S., Dontu, G., and Wicha, M.S. (2005). Mammary stem cells, self-renewal pathways, and carcinogenesis. Breast Cancer Res **7**, 86–95.

89. Weng, A.P., and Aster, J.C. (2004). Multiple niches for Notch in cancer: context is everything. Curr Opin Genet Dev **14**, 48–54.

90. Baron, M. (2003). An overview of the Notch signalling pathway. Semin Cell Dev Biol **14**, 113–119.

91. Wakamatsu, Y., Maynard, T.M., Jones, S.U., and Weston, J.A. (1999). NUMB localizes in the basal cortex of mitotic avian neuroepithelial cells and modulates neuronal differentiation by binding to NOTCH-1. Neuron **23**, 71–81.

92. Okano, H., Imai, T., and Okabe, M. (2002). Musashi: a translational regulator of cell fate. J Cell Sci **115**, 1355–1359.

93. Clarke, R.B., Anderson, E., Howell, A., and Potten, C.S. (2003). Regulation of human breast epithelial stem cells. Cell Prolif **36**(Suppl 1), 45–58.

94. Nishimura, S., Wakabayashi, N., Toyoda, K., Kashima, K., and Mitsufuji, S. (2003). Expression of Musashi-1 in human normal colon crypt cells: a possible stem cell marker of human colon epithelium. Dig Dis Sci **48**, 1523–1529.

95. Sakakibara, S., Imai, T., Hamaguchi, K., Okabe, M., Aruga, J., Nakajima, K., Yasutomi, D., Nagata, T., Kurihara, Y., Uesugi, S., Miyata, T., Ogawa, M., Mikoshiba, K., and Okano, H. (1996). Mouse-Musashi-1, a neural RNA-binding protein highly enriched in the mammalian CNS stem cell. Dev Biol **176**, 230–242.

96. Siddall, N.A., McLaughlin, E.A., Marriner, N.L., and Hime, G.R. (2006). The RNA-binding protein Musashi is required intrinsically to maintain stem cell identity. Proc Natl Acad Sci U S A **103**, 8402–8407.

97. Sugiyama-Nakagiri, Y., Akiyama, M., Shibata, S., Okano, H., and Shimizu, H. (2006). Expression of RNA-binding protein Musashi in hair follicle development and hair cycle progression. Am J Pathol **168**, 80–92.

98. McGill, M.A., and McGlade, C.J. (2003). Mammalian numb proteins promote Notch1 receptor ubiquitination and degradation of the Notch1 intracellular domain. J Biol Chem **278**, 23196–23203.

99. Yokota, N., Mainprize, T.G., Taylor, M.D., Kohata, T., Loreto, M., Ueda, S., Dura, W., Grajkowska, W., Kuo, J.S., and Rutka, J.T. (2004). Identification of differentially expressed and developmentally regulated genes in medulloblastoma using suppression subtraction hybridization. Oncogene **23**, 3444–3453.

100. Nakamura, M., Okano, H., Blendy, J.A., and Montell, C. (1994). Musashi, a neural RNA-binding protein required for Drosophila adult external sensory organ development. Neuron **13**, 67–81.

101. Sakatani, T., Kaneda, A., Iacobuzio-Donahue, C.A., Carter, M.G., de Boom, W.S., Okano, H., Ko, M.S., Ohlsson, R., Longo, D.L., and Feinberg, A.P. (2005). Loss of imprinting of IGF2 alters intestinal maturation and tumorigenesis in mice. Science **307**, 1976–1978.

102. Antic, D., and Keene, J.D. (1997). Embryonic lethal abnormal visual RNA-binding proteins involved in growth, differentiation, and posttranscriptional gene expression. Am J Hum Genet **61**, 273–278.

103. Okano, H.J., and Darnell, R.B. (1997). A hierarchy of Hu RNA binding proteins in developing and adult neurons. J Neurosci **17**, 3024–3037.

104. Ratti, A., Fallini, C., Cova, L., Fantozzi, R., Calzarossa, C., Zennaro, E., Pascale, A., Quattrone, A., and Silani, V. (2006). A role for the ELAV RNA-binding proteins in neural stem cells: stabilization of Msi1 mRNA. J Cell Sci **119**, 1442–1452.

105. Pascale, A., Amadio, M., Scapagnini, G., Lanni, C., Racchi, M., Provenzani, A., Govoni, S., Alkon, D.L., and Quattrone, A. (2005). Neuronal ELAV proteins enhance mRNA stability by a PKCalpha-dependent pathway. Proc Natl Acad Sci U S A **102**, 12065–12070.

106. Akamatsu, W., Fujihara, H., Mitsuhashi, T., Yano, M., Shibata, S., Hayakawa, Y., Okano, H.J., Sakakibara, S., Takano, H., Takano, T., Takahashi, T., Noda, T., and Okano, H. (2005). The RNA-binding protein HuD regulates neuronal cell identity and maturation. Proc Natl Acad Sci U S A **102**, 4625–4630.

107. Szabo, A., Dalmau, J., Manley, G., Rosenfeld, M., Wong, E., Henson, J., Posner, J.B., and Furneaux, H.M. (1991). HuD, a paraneoplastic encephalomyelitis antigen, contains RNA-binding domains and is homologous to Elav and Sex-lethal. Cell **67**, 325–333.

108. Imai, T., Tokunaga, A., Yoshida, T., Hashimoto, M., Mikoshiba, K., Weinmaster, G., Nakafuku, M., and Okano, H. (2001). The neural RNA-binding protein Musashi1 translationally regulates mammalian numb gene expression by interacting with its mRNA. Mol Cell Biol **21**, 3888–3900.

109. Okabe, M., Imai, T., Kurusu, M., Hiromi, Y., and Okano, H. (2001). Translational repression determines a neuronal potential in Drosophila asymmetric cell division. Nature **411**, 94–98.

110. Baonza, A., Murawsky, C.M., Travers, A.A., and Freeman, M. (2002). Pointed and Tramtrack69 establish an EGFR-dependent transcriptional switch to regulate mitosis. Nat Cell Biol **4**, 976–980.

111. Battelli, C., Nikopoulos, G.N., Mitchell, J.G., and Verdi, J.M. (2006). The RNA-binding protein Musashi-1 regulates neural development through the translational repression of p21WAF-1. Mol Cell Neurosci **31**, 85–96.

112. Li, F., Ackermann, E.J., Bennett, C.F., Rothermel, A.L., Plescia, J., Tognin, S., Villa, A., Marchisio, P.C., and Altieri, D.C. (1999). Pleiotropic cell-division defects and apoptosis induced by interference with survivin function. Nat Cell Biol **1**, 461–466.

113. Devgan, V., Nguyen, B.C., Oh, H., and Dotto, G.P. (2006). p21WAF1/Cip1 suppresses keratinocyte differentiation independently of the cell cycle through transcriptional up-regulation of the IGF-I gene. J Biol Chem **281**, 30463–30470.

114. Kaplan, D.D., Meigs, T.E., Kelly, P., and Casey, P.J. (2004). Identification of a role for beta-catenin in the establishment of a bipolar mitotic spindle. J Biol Chem **279**, 10829–10832.

115. Okano, H., Kawahara, H., Toriya, M., Nakao, K., Shibata, S., and Imai, T. (2005). Function of RNA-binding protein Musashi-1 in stem cells. Exp Cell Res **306**, 349–356.

116. Ayyanan, A., Civenni, G., Ciarloni, L., Morel, C., Mueller, N., Lefort, K., Mandinova, A., Raffoul, W., Fiche, M., Dotto, G.P., and Brisken, C. (2006). Increased Wnt signaling triggers oncogenic conversion of human breast epithelial cells by a Notch-dependent mechanism. Proc Natl Acad Sci U S A **103**, 3799–3804.

117. Reedijk, M., Odorcic, S., Chang, L., Zhang, H., Miller, N., McCready, D.R., Lockwood, G., and Egan, S.E. (2005). High-level coexpression of JAG1 and NOTCH1 is observed in human breast cancer and is associated with poor overall survival. Cancer Res **65**, 8530–8537.

118. Stylianou, S., Clarke, R.B., and Brennan, K. (2006). Aberrant activation of Notch signaling in human breast cancer. Cancer Res **66**, 1517–1525.

119. Pece, S., Serresi, M., Santolini, E., Capra, M., Hulleman, E., Galimberti, V., Zurrida, S., Maisonneuve, P., Viale, G., and Di Fiore, P.P. (2004). Loss of negative regulation by Numb over Notch is relevant to human breast carcinogenesis. J Cell Biol **167**, 215–221.

120. Hu, C., Dievart, A., Lupien, M., Calvo, E., Tremblay, G., and Jolicoeur, P. (2006). Overexpression of activated murine Notch1 and Notch3 in transgenic mice blocks mammary gland development and induces mammary tumors. Am J Pathol **168**, 973–990.

121. Jhappan, C., Gallahan, D., Stahle, C., Chu, E., Smith, G.H., Merlino, G., and Callahan, R. (1992). Expression of an activated Notch-related int-3 transgene interferes with cell differentiation and induces neoplastic transformation in mammary and salivary glands. Genes Dev **6**, 345–355.

122. Fitzgerald, K., Harrington, A., and Leder, P. (2000). Ras pathway signals are required for notch-mediated oncogenesis. Oncogene **19**, 4191–4198.

123. Gonsalves, F.C., and Weisblat, D.A. (2007). MAPK regulation of maternal and zygotic Notch transcript stability in early development. Proc Natl Acad Sci U S A **104**, 531–536.

124. Stockhausen, M.T., Sjolund, J., and Axelson, H. (2005). Regulation of the Notch target gene Hes-1 by TGFalpha induced Ras/MAPK signaling in human neuroblastoma cells. Exp Cell Res **310**, 218–228.

125. Klinakis, A., Szabolcs, M., Politi, K., Kiaris, H., Artavanis-Tsakonas, S., and Efstratiadis, A. (2006). Myc is a Notch1 transcriptional target and a requisite for Notch1-induced mammary tumorigenesis in mice. Proc Natl Acad Sci U S A **103**, 9262–9267.

126. Kiaris, H., Politi, K., Grimm, L.M., Szabolcs, M., Fisher, P., Efstratiadis, A., and Artavanis-Tsakonas, S. (2004). Modulation of notch signaling elicits signature tumors and inhibits hras1-induced oncogenesis in the mouse mammary epithelium. Am J Pathol **165**, 695–705.

127. Dontu, G., Abdallah, W.M., Foley, J.M., Jackson, K.W., Clarke, M.F., Kawamura, M.J., and Wicha, M.S. (2003). In vitro propagation and transcriptional profiling of human mammary stem/progenitor cells. Genes Dev **17**, 1253–1270.

128. Smith, G.H., Sharp, R., Kordon, E.C., Jhappan, C., and Merlino, G. (1995). Transforming growth factor-alpha promotes mammary tumorigenesis through selective survival and growth of secretory epithelial cells. Am J Pathol **147**, 1081–1096.

129. Clarke, R.B., Spence, K., Anderson, E., Howell, A., Okano, H., and Potten, C.S. (2005). A putative human breast stem cell population is enriched for steroid receptor-positive cells. Dev Biol **277**, 443–456.

130. Clarke, R.B. (2005). Isolation and characterization of human mammary stem cells. Cell Prolif **38**, 375–386.

131. Mizumoto, K., and Sawa, H. (2007). Two betas or not two betas: regulation of asymmetric division by beta-catenin. Trends Cell Biol **17**, 465–473.

132. Ozdamar, B., Bose, R., Barrios-Rodiles, M., Wang, H.R., Zhang, Y., and Wrana, J.L. (2005). Regulation of the polarity protein Par6 by TGFbeta receptors controls epithelial cell plasticity. Science **307**, 1603–1609.

133. Bose, R., and Wrana, J.L. (2006). Regulation of Par6 by extracellular signals. Curr Opin Cell Biol **18**, 206–212.

134. Perez-Moreno, M., Jamora, C., and Fuchs, E. (2003). Sticky business: orchestrating cellular signals at adherens junctions. Cell **112**, 535–548.

135. Meshorer, E., and Gruenbaum, Y. (2008). Gone with the Wnt/Notch: stem cells in laminopathies, progeria, and aging. J Cell Biol **181**, 9–13.

136. Brack, A.S., Conboy, M.J., Roy, S., Lee, M., Kuo, C.J., Keller, C., and Rando, T.A. (2007). Increased Wnt signaling during aging alters muscle stem cell fate and increases fibrosis. Science **317**, 807–810.

137. Liu, H., Fergusson, M.M., Castilho, R.M., Liu, J., Cao, L., Chen, J., Malide, D., Rovira, I.I., Schimel, D., Kuo, C.J., Gutkind, J.S., Hwang, P.M., and Finkel, T. (2007). Augmented Wnt signaling in a mammalian model of accelerated aging. Science **317**, 803–806.

138. Scaffidi, P., and Misteli, T. (2006). Lamin A-dependent nuclear defects in human aging. Science **312**, 1059–1063.

139. Cao, K., Capell, B.C., Erdos, M.R., Djabali, K., and Collins, F.S. (2007). A lamin A protein isoform overexpressed in Hutchinson-Gilford progeria syndrome interferes with mitosis in progeria and normal cells. Proc Natl Acad Sci U S A **104**, 4949–4954.

140. Liang, Y., and Van Zant, G. (2008). Aging stem cells, latexin, and longevity. Exp Cell Res **314**, 1962–1972.

141. Oakley, E.J., and Van Zant, G. (2007). Unraveling the complex regulation of stem cells: implications for aging and cancer. Leukemia **21**, 612–621.

142. Shcheprova, Z., Baldi, S., Frei, S.B., Gonnet, G., and Barral, Y. (2008). A mechanism for asymmetric segregation of age during yeast budding. Nature **454**, 728–734.

143. Fuentealba, L.C., Eivers, E., Geissert, D., Taelman, V., and De Robertis, E.M. (2008). Asymmetric mitosis: unequal segregation of proteins destined for degradation. Proc Natl Acad Sci U S A **105**, 7732–7737.
144. Coumailleau, F., and Gonzalez-Gaitan, M. (2008). From endocytosis to tumors through asymmetric cell division of stem cells. Curr Opin Cell Biol **20**, 462–469.
145. Erik Rusten, T.E., Filimonenko, M., Rodahl, L.M., Stenmark, H., and Simonsen, A. (2008). ESCRTing autophagic clearance of aggregating proteins. Autophagy **4**, 233–236.
146. Rusten, T.E., and Simonsen, A. (2008). ESCRT functions in autophagy and associated disease. Cell Cycle **7**, 1166–1172.
147. Moberg, K.H., Schelble, S., Burdick, S.K., and Hariharan, I.K. (2005). Mutations in erupted, the Drosophila ortholog of mammalian tumor susceptibility gene 101, elicit non-cell-autonomous overgrowth. Dev Cell **9**, 699–710.
148. Carlton, J.G., and Martin-Serrano, J. (2007). Parallels between cytokinesis and retroviral budding: a role for the ESCRT machinery. Science **316**, 1908–1912.
149. Carlton, J.G., Agromayor, M., and Martin-Serrano, J. (2008). Differential requirements for Alix and ESCRT-III in cytokinesis and HIV-1 release. Proc Natl Acad Sci U S A **105**, 10541–10546.
150. Easwaran, V., Pishvaian, M., Salimuddin, and Byers, S. (1999). Cross-regulation of beta-catenin-LEF/TCF and retinoid signaling pathways. Curr Biol **9**, 1415–1418.
151. Palmer, H.G., Gonzalez-Sancho, J.M., Espada, J., Berciano, M.T., Puig, I., Baulida, J., Quintanilla, M., Cano, A., de Herreros, A.G., Lafarga, M., and Munoz, A. (2001). Vitamin D(3) promotes the differentiation of colon carcinoma cells by the induction of E-cadherin and the inhibition of beta-catenin signaling. J Cell Biol **154**, 369–387.
152. Shah, S., Hecht, A., Pestell, R., and Byers, S.W. (2003). Trans-repression of beta-catenin activity by nuclear receptors. J Biol Chem **278**, 48137–48145.
153. Shah, S., Islam, M.N., Dakshanamurthy, S., Rizvi, I., Rao, M., Herrell, R., Zinser, G., Valrance, M., Aranda, A., Moras, D., Norman, A., Welsh, J., and Byers, S.W. (2006). The molecular basis of vitamin D receptor and beta-catenin crossregulation. Mol Cell **21**, 799–809.
154. Nowak, J.A., Polak, L., Pasolli, H.A., and Fuchs, E. (2008). Hair follicle stem cells are specified and function in early skin morphogenesis. Cell Stem Cell **3**, 33–43.
155. Cirillo, L.A., and Zaret, K.S. (2007). Specific interactions of the wing domains of FOXA1 transcription factor with DNA. J Mol Biol **366**, 720–724.

Index

Printed in the United States
by Baker & Taylor Publisher Services